Edingow
and
Glasburgh

EDINGOW
and
GLASBURGH

by
Peter Terrell

illustrated by
Elfreda Crehan

Published 2018 by Lexus Ltd
47 Broad Street, Glasgow G40 2QW

Compiled by Peter Terrell
Cover design, page design and illustrations by Elfreda Crehan

The cover uses John Slezer's
The North Prospect of the City of Edenburgh (c. 1690) and
The Prospect of ye Town of Glasgow from ye North East
(1693) both reproduced by permission of the
National Library of Scotland, with details from each
occurring throughout the book.

www.lexusforlanguages.co.uk

British Library Cataloguing in Publication Data
A catalogue record for this book is available from the British
Library.
ISBN: 978-1-904737-490

Printed and bound in Europe by PULSIO SARL

Edinburgh Castle high above the Nor' Loch, which is now Princes Street Gardens
© National Library of Scotland

INTRODUCTION

WHAT is Gow? What is Edin? Or Glas? Burgh is more familiar. What do these bits of placenames mean and why are places called what they are called? Do these names and bits of names mean anything at all? What lies behind it all?

Most of the time we take names for granted, a major exception to this being when choosing a name for a newborn baby or, perhaps, when setting up a new company or building a team. Is there something special about names as such? A name seems to be a different type of word. One of the characteristics of a name, as opposed to other types of words, is that it may be used and heard in such a way that its reference, what is named, is clearly understood while at the same time the name itself remains devoid of meaning.

Most travellers to Scotland, and natives too, will be struck by the Auchenshuggles and Restalrigs that confront them. What language is this? Where do these names come from? Do they really have no meaning?

This book takes 200 placenames, 100 from each of two Scottish cities, and examines them. The names can be like time capsules. When you open them up, dissect them, and look beneath the surface of something whose function is simply to refer to and identify a place then what is revealed can, in many instances, in fact be a meaning too. In some cases the examination will reveal glimpses of our ancestors' lives, of old practices, activities and deeds, glimpses of worlds that have gone.

There are many different reasons for giving a place a name. The name may have, or may once have had, meaning if, for example, it describes a local feature of the landscape: a hill, a river bend, a marsh, a fort, a religious

site. Or a placename may have meaning if it describes
an activity that characterized a district or site: a storage
place for cow dung, a smithy, a mill. Or a name may have
no actual meaning but just a reference, particularly if the
name described the ownership of a property.

Some examples of reasons for naming

Proud owners of farmsteads which they set up and ran
amid surrounding marshland used their own personal
name to name the farm, which grew to become a hamlet,
then a village, then a city district (Clermiston, Warriston).
Or more probably it was the people who lived round about
who gave the farm its name, the name being more likely to
have taken root if collective use was behind it.

Straightforward descriptions of a significant feature of
the landscape are a common reason for naming (Cammo,
Camlachie).

Names may come from main activities carried out in an
area (Lawnmarket, Candleriggs).

References to a powerful or significant person are often
embedded in placenames (Corstorphine, Cardonald).

Nicknames may have taken on permanent status, quite
possibly for an area or place that didn't really have a
generally agreed name before (Little France). Of course, the
point of the nickname becomes lost with the passage of time.

Nostalgia can also be a reason for giving a place a name,
nostalgia for a place left forever behind (Moredun).
Names like this may at first seem to be fakes if assumed
to be descriptions of the landscape or of manmade
structures, until it is discovered that the name was in
fact transplanted from a place where it did match the
landscape or describe a structure. These are names that
have meaning but no actual local reference.

Closely related to nostalgic naming is naming by reason
of some event of great personal and possibly life-changing
significance (Portobello).

Some names may be cover-ups, distortions of a true and factual description in order to hide an unwelcome past or present truth (Fountainbridge).

Religion has played a very big role in naming, in particular the once newly spreading Christian religion (Govan, Inchmickery). Also in later centuries there was a predilection for naming a property or a farm by using a biblical reference (Joppa, Bellahouston).

What methods are used to find a name's meaning?

Older texts, records and maps are the tools that can be used in examining names. These show older written forms of names, often with a large variety of spellings. When looking at older maps, for example, it soon becomes apparent that either the map-makers were chronically poor spellers or most probably it would be more accurate to say that their informants, the people who spoke, or perhaps wrote down, the names of the places being mapped, provided this naming information in a mixture of accents and in differing spellings. And there is nothing surprising in this. Standard forms were few and far between. But does this play any role in the investigation of the possible meaning or reference of placenames? It may. And it may not. For example, the fact that a late 17th century map of Midlothian shows Leith as Lieth is of no importance. However the fact that one occurrence of the older name forms of Baberton contains the letter l – Balberton – throws the brightest of lights on the origin of this name. Here we are not looking at just a mispelling or a spelling variant but at a form of the name that illuminates the component parts, parts that time and usage have gradually worn away.

Are there unknowables? Where a name is agreed to be nothing more than a reference, the answer to this is no. Take, for example, the Edinburgh district of Lauriston. The name has a reference; it was, when it first came about, the home and farmstead of a man called Laurence, very probably a man of Norman descent. Exactly who this Laurence was and what sort of person he was are things that cannot be known. But the reference of the name is not unknowable. In cases like this there was and is no meaning over and above the reference.

There are other placenames, which are descriptive in nature and which can be assumed to have had meaning, and which may sometimes, if the lie of the land has not changed or the course and width of a river has not altered, still be seen to match that description, once the description has been deciphered. There are, however, more than a few instances of descriptive naming where two or more explanations of the meaning of the name are put forward. In such instances it all comes down to an evaluation of plausibility and a closer examination of the surrounding facts. Two or more naming origins are, in general, not as good as one. But two are better than none.

Some of the most elusive and mysterious names (Edinburgh) are names where it is hard, or impossible, to establish whether it is meaning or simple reference that is to be looked for and accepted as the origin. And in that sense of unknowable there are indeed unknowables.

LANGUAGES

There are several languages at work in the creation of these two cities' placenames. The oldest is called Cumbric and is the precursor of modern Welsh, and many of the Cumbric terms occurring in placenames are still recognizable in Welsh. Then there is Scots Gaelic, some Gaelic words showing an obvious relationship to Cumbric. Old English is part of the mixture too, especially in endings such as –ton or –wick. Scots words are another part of the mix with terms like shaw, braid, law. Nordic words and names were also material for placenames.

The 100 Glasgow names examined have a higher proportion of Gaelic in their origins. The 100 Edinburgh names show the marks of greater influence from the Anglian language spoken to the south, as well as some French and Norse in personal names.

PRONUNCIATION

This book gives a pronunciation guide for placenames where the spoken form is not obvious from the written form. And where relevant, and possibly useful, a pronunciation guide is given for related or original Gaelic forms or components of names. There are times when this may illuminate but, more often than not, knowing the (modern) Gaelic pronunciation of words in a name only serves to emphasize how far the modern pronunciation has moved away from how a name may originally have been spoken. Intervening languages, particularly English and Scots, will have adapted the Gaelic. Other languages in placenames will have undergone similar adjustments, since the speakers of the predominant languages of Scots and English could not or would not get their tongues around the originals.

It is not uncommon for a placename to be made up of elements deriving from more than a single language and for the emerging form of the name to have taken on the pronunciation and written form of just one of these languages – usually Scots or English, although there are instances (Sciennes) where most of an older written form stubbornly survives.

The present author acknowledges the work of others who have investigated the history and origins of Scottish placenames and is indebted to them for their discoveries and insights. *Edingow and Glasburgh* may add to what can be said about just two hundred of the names that populate two great cities of this nation.

PT
Glasgow, 2018

EDINBURGH

The view from Calton Hill with the spire of St Giles' Cathedral to the
right and, further down the Royal Mile, the spire of the Tron Church
© National Library of Scotland

Arthur's Seat and Salisbury Crags
© National Library of Scotland

Arthur's Seat

The name of this section of Edinburgh's remaining chunks of volcanic rock (thought to have been formed some 350 million years ago) is generally believed to stem from the ancient Britons' habit of commemorating their legendary King Arthur by associating significant features of their landscape with him. But this is a naming origin that is more likely to stem from the Middle Ages, with a touch of romanticism, than from more distant times. The earliest written record which refers to Arthurian legend dates from 1508; this has it as Arthurissete. Holyrood charters from the 12th century, which specifically concern the division of land hereabouts, call the hill Craggenmarf and make no mention at all of Arthur. The name Craggenmarf can be seen through either Cumbric or Gaelic as meaning hill of the dead.

Baberton

In the 1300s present-day Baberton was recorded as both Kibabirtone and Kilbabertone. The Ki(l), which subsequently disappeared from the name, is possibly Gaelic *cill* meaning church or, more often, a holy man's cell. It is likely that the remaining and current name is also of Gaelic origin, even though this is not immediately obvious. Blaeu's map of 1654 shows a tower here by the name of Balbertoun. And that letter l in the name of Balbertoun is a clue pointing to a name originating with the very widely used Gaelic *baile* (homestead). Although at this stage the –toun is showing a clear pull of the Scots, the modern –berton is likely to have emerged from a name that had undergone the same sort of mutation as the –barton in Dumbarton, which is to say that it means 'of the Britons'. This then was a place once inhabited by Britons, probably from Strathclyde, or maybe people descended from the old Votadini tribe (the Iron-Age people that inhabited these parts up to the years of the Roman presence); at any rate, they were people seen as a distinct group by local Gaelic speakers. The name eventually naturalized, lost its Gaelic aspects and changed to something resembling, on the surface, other names in the more familiar English or Scottish –ton.

BALERNO

This name comes down from the Gaelic. There are two elements: a *baile* [baluh] can be anything from a small town to a farm, a settlement, a place where humans are active; and *àirne* is the sloe (berry or bush), also known as the blackthorn. Sloe berries can be used for making a kind of gin; and their prickly bushes make a reasonable barrier (if you happened to want to mark out a farmstead). Gaelic-origin names are not particularly common in and around Edinburgh, but Balerno is a notable exception.

BANGHOLM

[pronounced: bang-uhm]

Holm is Scots for the low-lying, rich land beside a river. The Bang– is likely to be a flattened-out version of bank, which can be a peat bank, a place where peat is cut. This was originally the name of a farm.

Beeslack

A *slack* is a Scots word for a hollow or lower ground between hills; it can also mean a boggy hollow. Beeslack is a slack renowned for its bees as well as providing an irresistible name for Beeslack High School.

Bingham

A *bing* in Scots is a slag-heap. In the 17th and 18th centuries the neighbouring district of Duddingston was a centre of coal mining and of salt mining. And Bingham, it would seem, got the slag.

Blinkbonny

It might be Scots with the Blink– being a glimpse or a quickish view. And –bonny is bonnie or fair. But why is the adjective put after the noun as though it were Gaelic or French? You have to suspect the influence of a generally emotive aspect to the naming. Blinkbonny is used for several farms which are south-facing.

Bonaly

[pronounced: bon-ah-lee]

The Gaelic words *bonn* (bottom) and *àth* (ford) are at the root of many Scottish placenames (Bonar Bridge, Bonawe). And they very likely play a role here too. Early (14th century) recorded forms of

the name are Bounaylin, Benhathelyn, Bonalyn.
The endings –lin or –lyn point to Gaelic *linn*
(pool). But the final n of this ending drops away
after the 14th century. In 1654 Blaeu's map has
it as Bonely. But, other than that, the second
syllable –a– is consistently present in the name
variations; and carries the stress. So this is likely
to be the pool at the low ford. The old village of
Bonaly stood at the confluence of two burns (before
being demolished by the landowner to make way
for Bonaly Tower), adding weight to the watery
interpretation.

BONNINGTON

Centuries ago this suburb was a small village
clustered around a ford and water mills on the
Water of Leith. In the 12th century the land here
was known as Bonnytoun, in simple recognition,
you might think, of its charm. Well no, the bonny
here refers not to appearance but to ownership and
goes back to a man of Norse extraction (or bearing
a Norse name) called Bondi who owned a farm
or homestead or *tún* here. The –ing– in names
with this structure usually carries the sense of 'the
people of'. So this was Bondi's people's place.

BONNYRIGG

Bonnyrigg has a curious naming history. Around
1750 it was mapped as Bonnebrig (fine bridge).
Around 1763 this had become Bannockrigg, which
was – if we take the Bannock– to parallel the
Bannock as in Bannockburn and to derive from a

Cumbric word *bannauc* (peaked hill) – the ridge
of/by the peaked hill. In 1817 the name had become
Bonny Ridge. And it was left to the Ordnance
Survey to standardize the name into its current
form. But the shift from ridge to rigg may point to
a Scots sense of rigg, which is a strip of arable land.
And *bonnage* was a Scots word for the labour and
duty owed by a tenant farmer to the owner of the
land he or she worked.

BRAID

Although *braid* is Scots for broad and the name of
the Braid Burn may tempt you into thinking this
is a broad burn, this Edinburgh placename can't
really be seen as deriving just from an adjective
meaning broad. One theory traces the name of
Braid back to a Flemish knight who came north
with King David I in the 12th century, a knight by
the name of Richard de Brad or Richard de Breda,
to whom land was granted in return for service to
the king. But a more likely derivation takes us back
to an old Gaelic word *bràid* [braj] which can mean
mountainous country and which is possibly an
older form of the Gaelic word *bràghad* [brah-uht]
(upper part, uplands). The same root is present in
other placenames such as Breadalbane or Braemar
and, in fact, in the Scots *braes* for hills. The Braid
Hills are really just the hilly hills. As ever, an
anglicizing pronunciation of the written form has
superseded the original.

BROUGHTON

[pronounced: broh-tun]

This is a name very commonly found throughout England though it is less common in Scotland. The Brough– derives from the old English word *bróc* (meaning brook) and the –ton is for the farm(s) or settlement by it, the *tún*.

BRUNTSFIELD

In the 14th century the land here was known as *Brounysfelde* or Brown's Fields. These were the fields belonging to one Richard Broun of Boroumore, the first owner of the property that came to be known as Bruntsfield House. Adair's map of 1682 has it as Brunsfield – minus that intrusive t, which isn't pronounced, so we hear, by the locals.

BUGHTLIN

[pronounced: bootlin]

'ught' without a preceding o or a is an unusual grouping of letters in modern English and is likely to result from an attempt at transcription of an earlier name for which no standard written form existed. Modern Welsh can help here if we accept that Welsh is a living descendant of the Cumbric language once spoken here. Welsh *llyn* is pond and *buwch* is cow. This would be a good reason for naming a place; and there are other cases in Scotland (such as Bowling on the Clyde) where the regular need to take cattle to water results in a place acquiring a name. Contrasting with Cumbric cows, however, in Scots a *boucht*, *bought* or *bucht* is a sheep-fold. Whether sheep or cows or both, this was a fold or pen by a pond.

BURDIEHOUSE

One existing explanation of this name is that there is a connection here with the French city of Bordeaux. How so?

The Reverend Dr John Jamieson's Dictionary of the Scottish Language, first published in 1808, says that Burdiehouse seems to be an older pronunciation of the name of Bordeaux in France, a Scots written form of which was Burdeouss, and which, he says, 'was probably aspirated by the vulgar in pronunciation' (ie –ouss became –house). (His interest, incidentally, was not the placename itself but the use of the defunct old Scots slang expression 'gang to Burdiehouse!' which meant something like 'you should be ashamed of yourself!').

So a theory developed that this name was given with reference to the sudden influx of French courtiers and attendants of Mary Queen of Scots, who came back to Scotland from France in the year 1561. Mary stayed at Craigmillar Castle, only a short distance away, and the area there became known as Little France, so a French origin for Burdiehouse was seen as plausible. Were some of the courtiers from Bordeaux?

We cannot know. Or were French people from Bordeaux perhaps already here before Mary's arrival?

We cannot know. Did the locals, as suggested by Jamieson's description of the slang phrase, look on the French incomers with disapproval?

None of this will really stand up in court. For one thing, any speaker of Scots or English will tell you that it is far more likely for an h to be dropped than it is for an h to be inserted. So to say, with Dr Jamieson, that Burdehouss was a spoken variant of Burdeouss does seem like planting false evidence.

Evidence from old map-makers is enlightening (although still puzzling). Adair's 1682 map of Midlothian shows the name as Burdeaux, lying just southwest of Sudhouse. Is this not an argument in favour of the Bordeaux theory? But then on the next Adair map, in 1735, in which the spelling has been generally tightened up, Burdeaux has become Burdehouse and is mapped lying south of, but not quite so far southeast of, Sudhouse, neither of these two placenames being shown on this later map with an accompanying drawing of a mansion or small castle, thus indicating that they were of no significant size (although the drawings were there on the 1682 map). It looks as though the Adair map-maker had had a change of mind, that he had received new information.

Sudhouse offers a clue. Sudhouse is a placename that has now vanished. But what was it? *Sud* is a variant spelling of Scots *sid*, defined as 'the inner husks of oats after grinding'. So Sudhouse and its neighbouring Burdehouse were most probably not houses for habitation at all, but rather buildings for storage. Sudhouse for storing grain and Burdehouse was a storage house for produce from the fields for the laird's or landowner's table (for his *bord*). This fits a pattern of similarly derived names, such as Boreland in Dumfries and Galloway, which comes from *bordland,* a now obsolete legal term for the land kept by a landowner for supplying the needs of his own pantry and table.

Calton: the Calton Hill

The name of this hill was first recorded in 1456 as Cragingalt, a name which could be understood via Welsh as a Cumbric name meaning rock with wooded slopes. But a name change occurred. 16th century records show a change to Cauldtoun and the current name of Calton appeared in the early 18th century. Some have attributed a Gaelic meaning to the name, putting it on a par with the Glaswegian Calton, that name coming from the Gaelic *calltainn* (hazels).

Cammo

It is thought that this name comes from, or is related to, the Gaelic *cam* meaning a bend. The second part, the –mo, can be seen as a truncated version of the Gaelic ending *–ach*, which is pronounced with a short o sound and which, as well as meaning field, is also used to mean something like 'place of'. On the earlier of Adair's maps it is given as Cammuck (1682), closer to the Gaelic pronunciation; it has changed to its present form of Cammo on the later Adair map (1735). It's not unlikely that the bend in question is the nearby bend in the River Almond, a significant feature of the landscape in that it would have provided a place where boats could be pulled up onto the shore.

CANONGATE

The Canongate, as depicted above running from right to left, was not a gate but the street (Old English *geata*) of the canons, the canons being the Augustinians at Holyrood Abbey, founded by King David I of Scotland in 1128. King David gave the canons the right to establish their own independent burgh. And right up until 1856 Canongate and Edinburgh were to remain separate burghs.

CLERMISTON

The name comes down from an old personal name which has been squashed down to a more manageable size from the first recorded *tún* or farmstead of one Clerribaldi (1288).

Coates

Coates is/are just cottages or houses, from Old English *cote*.

Colinton

This was the *tún*, the farm, of a man called Colban, and the place was known back in 1319 as Colbanestoun. The letter b had disappeared, at least in written records, by 1488 and, although a g appeared after the –in– (Colingstoune) in the 16th century, it did not stay that long, and the modern form held fast.

Comely Bank

Comely Bank was new as a residential area in the early part of the 19th century and its name simply reflects the opinion that this was a fine-looking, highly desirable place to live. Those who came to live here, like the author Thomas Carlyle, would find themselves at some remove from the crowds, noise, smells and bustle of the wynds and closes of the Old Town up the hill.

Comiston

The name has been worn down over the centuries but can be found as Colmanstone in the 14th century. The –ton is a common ending, signifying a homestead or farm. St Colman was a 7th century Irish saint who trained on Iona and later became Bishop of Lindisfarne (AD661-664). The owner of this farmstead was named after him.

CORSTORPHINE

There is a connection here with Thorfinn the Mighty. What we know about Thorfinn the Mighty comes down from sagas, mainly from the Orkneyinga Saga, an Icelandic account of the lives of the Norsemen of Orkney, written in the first half of the 13th century. Thorfinn was an 11th century Earl of Orkney. The Saga describes him as being unusually tall and strong, black-haired, big-nosed, bushy-browed and belligerent. It is also recounted that he was a devout Christian and that he went on a pilgrimage to Rome in or around the year 1048. We may think it unlikely that his journey through the countryside and towns would have gone unnoticed.

One theory behind the name Corstorphine is that this is the place of Thorfinn's cross, the Cors– element of the name being a mutation of cross. Against this it is said that this was not Thorfinn's cross but his crossing. But his crossing of what? Well, old maps of the area show the existence of lochs at Corstorphine (drained in 1650 and previously stretching as far as Haymarket) and also at Gogar (drained around 1766). There was boggy and treacherous land all about and a party led by Thorfinn the Mighty would have good cause to celebrate their crossing through this district into old Edinburgh.

So *Cors* is crossing? Well, maybe not. This name element, Cors–, may have no connection with crossing. Cors can still be found in modern

Welsh, a language related to the Cumbric once spoken here. It means boggy ground. This then is Thorfinn's bog. Who knows what scenes unfolded here!

A further remnant of the district of Corstorphine's history is to be found in the oddly named Lampacre Road. This derives its name from the acre of dry land around Corstorphine parish church from the steeple of which a lamp used to be hung to act as a beacon for those, in the years after Thorfinn, heading across the bogs.

COWGATE

The Cow– refers uncomplicatedly to cows and the –gate to the Old English word *geata* which means road. On market days this was the cows' way in and (possibly) out.

CRAIGENTINNY

The first part of this name is from Gaelic *creag* (rock). –entinny is very likely to be from the Gaelic *an teine* [uhn chenuh] (of the fire), said fire being most probably a beacon fire. If that is the origin of the name, then this is a clear case of naming without reference to actual location. Craigentinny House (formerly Craigentinny Castle) was built in 1604 in flat and unrelieved moorland. Also the fact that an anglicized Gaelic name was chosen (as opposed to the actual Gaelic) is a sure sign that the owner simply liked the sound of this name or that it had some special significance for him and his family. Naming by whim.

Thorfinn the Mighty and his party, in or just before the year 1048, having travelled down from Orkney, now laboriously make their way across this treacherous boggy land before reaching Edinburgh, where they may well have stopped to recover. The rest of their pilgrimage to Rome cannot have been much more arduous than this.

The old loch at Corstorphine was drained in 1650 and the district became less boggy. The old parish church here, built on a patch of dry ground, used to hang a lamp from its steeple to guide travellers to safer ground. Corstorphine Hill can be seen through the trees.

CRAIGLOCKHART

This is the crag or rock (Scots *craig*) belonging to
a man called (in the original spelling of his name)
Loccard. The first record of this dates from the
year 1278. Ownership of a crag can be seen as
desirable if the crag serves the purpose of a ready-
made foundation for building works. The Loccards
became the Lockharts and built Craiglockhart
Castle here in the 15th century. The district takes
its name from the castle.

CRAIGMILLAR

Gaelic *creag* is a rock. The –millar can be accounted
for as Gaelic *maol* [murl] (bare, bleak) and *àrd*
[arsht] (high). This happens to sound like an
excellent description of the Castle Rock on which
Edinburgh Castle sits. But this name is that of
another castle, the impressive ruins of which stand
just a short stretch south of the modern estate that
bears the castle's name, Craigmillar. The castle
started as a tower house towards the end of the 14th
century and was expanded and added to over the
years. Mary Queen of Scots stayed at Craigmillar
on more than one occasion in the 1560s, finding the
place a safe but temporary refuge.

CRAMOND

This name comes from the Cumbric *caer* (fort) and
the name of the River Almond (Almond itself being
from a yet older term for water and probably cognate
with the Gaelic for river: *abhainn* [ah-vuhn]). The
fort in question is that which the Romans built here

in or around the year AD140 and which served to guard the supply port for the Roman army and personnel in Scotland. The Romans abandoned the fort around AD170, then briefly reinforced and reoccupied it around AD208, when it was again used as a fortified base for the campaigns of the Emperor Septimius Severus in what was the last attempt to bring Caledonia under the sway of the Roman Empire. Severus died in AD211 and, in the fullness of time, the fort became the property of the locals, the Votadini, who gave it its name.

CREWE TOLL

The name Crewe (which Adair's 1682 map has as Crou, retaining the pronunciation) probably comes from an old Cumbric name, which can be seen in Welsh *cryw* (stepping stones). The stepping stones would very likely have been a means of crossing over the nearby Wardie Burn.

CURRIE

Although Gaelic was never the main language in these parts, there are pockets of Gaelic-based names, particularly towards the west of the city. This name appears to have a Gaelic origin in the word *currach* [kooruhK] meaning boggy land.

DALKEITH

This was the field by the wood, from Cumbric *dol* (field) and *coed* (wood). Gaelic has similar words in *dail* (field) and an old, now defunct, Gaelic term *coit* (wood).

Comings and goings at the Roman fort built around AD140 to man and guard their supply port at Cramond, where the River Almond flows into the Firth of Forth.

Present-day Cramond, a peaceful and picturesque suburb, or village, with moorings at the river mouth. Traces of the old Roman presence still exist in outlines of the fort's foundations.

DALRY

This is a name with a Gaelic origin; but it could be one of three origins. Gaelic *dail* [dal] (field) and *fhraoich* [rur-eeK] (heather). Or the –ry element (which takes the stress) may be from old Gaelic *ruighe* [ree-yuh] (lower slope). Or again the –ry element may be from Gaelic *righ* [ree-yuh] meaning king.

DANDERHALL

The –hall is certain to come from Scots *haugh* (a not uncommon Scottish sound shift), a *haugh* being level ground beside a watercourse. *Dander* is also a Scots word: the cinders from a smithy.

DEAN VILLAGE

The Dean here was not a person but a valley. Old English *denu* or *dene* as well as Scots *den* both refer to a narrow valley, most appropriately for this spot.

DREGHORN

The ending –horn is most likely from an Old English ending *ærn*, which simply means place. And the Dreg– is a time-warped remnant of Old English *drygge*, which means dry. In bog-surrounded old Edinburgh a dry place deserved recognition.

DRUMBRAE

From Gaelic *druim* [droo-im] (ridge) and Scots *brae* (which itself goes back to Gaelic *bràigh*). So this is Hill Ridge.

DRYLAW

A *law* in Scots is a hill, especially a rounded hill.
Singling out the dryness of it may well have been
a reference to the lack of a stream or source of
water on it. Or conversely it may be testimony to a
welcome absence of bogginess. The district takes it
name from Drylaw House, tucked away up on the
hill.

DUDDINGSTON

This name is thought to go back to the name of an
Anglo-Norman knight called Dodin de Berwic who
came into possession of the land here and referred
to himself, rather grandly, in or around 1150, as
Dodin of Dodinestoun. Local speech and local
record-keepers picked up the Dodinestoun and
modified it down the centuries.

DUMBIEDYKES

In the 1760s one Thomas Braidwood established
his Academy for the Deaf and Dumb in this area
of Edinburgh. The building in which he worked
and achieved remarkable successes with those
entrusted to his care then acquired the nickname
locally of Dumbie House (being pronounced, of
course, without a b).

Dykes are walls or fences, as well, sometimes, as
the ground enclosed by these. So the nickname
for Thomas Braidwood's famous academy was
soon extended to the whole district in which he
worked. And the name became established and

stuck, even though Braidwood himself closed down his Edinburgh academy and moved to London in 1783, to start a new academy. Dumbie House was demolished in 1939.

EDINBURGH

Two enquiring minds are sitting in an Edinburgh pub.

Prof. A.   Well, Raymond, it's a strange thing, is it not, that while most big cities can be fairly certain about the origin or meaning of their names, we have to admit that our capital, Edinburgh, poses a mystery.

Dr J.   Well, Charles, that is true only if you are not prepared to accept that the Edin in Edinburgh comes from the same old word that we, and other scholars, accept for other Scottish Edins, like Edinbane.

Prof. A.   You mean Gaelic *aodann*?

Dr J.   Yes. And if not directly Gaelic *aodann*, then, as with many other Gaelic words, we can take it that there was a related and earlier Brittonic or Cumbric word, also meaning face or rock face. And all these other instances of Edins relate to a rock face. You'll agree, I take it, that we have a pretty noticeable rock face here in Edinburgh.

Prof. A.	We do indeed. And more than one. But I think you might be jumping to conclusions.
Dr J.	How so?
Prof. A.	Well, tell me how do you explain the background to the name Carriden, a place not too far away from Edinburgh?
Dr J.	Well, the Carr is clearly from Cumbric *caer*, meaning fortress. And, as you know, it is still the Welsh for fortress. No doubt about that bit. Carriden is where the Romans ended the Antonine Wall with a fort.
Prof. A.	Agreed. And the iden is the same as the Edin in Edinburgh. But there's no rock face or even a hill face in or near Carriden.
Dr J.	So?
Prof. A.	So the iden here is much more likely to refer to the name of an area. The area where the *caer* is. Or was.
Dr J.	An area whose name has only survived in two placenames?
Prof. A.	Why not?
Dr J.	Hmm. You could as well say it's from Welsh *caer+ rhedyn*.
Prof. A.	You've got me there. Didn't know you were such an expert on the Welsh. What's *rhedyn*?

Dr J. Ferns. Fortress in the ferns. Which is probably what an old wooden fort looked like years after the Romans left.

Prof. A. Suppose so.

Dr J. Let's start again. Let's look at the possibilities. What do we have to go on? Where do we start?

Prof. A. With the usual suspects.

Dr J. OK. The main sources of a placename are going to be found in three types of reference: the name of an owner or dominant person; a description of an activity carried out in the place; a description of a relevant and significant natural feature. Agreed?

Prof. A. Agreed.

Dr J. Well, as far as naming activities is concerned, I don't think that Edin is going to take us anywhere.

Prof. A. Nor do I. What about people?

Dr J. The first known form of the name Edinburgh is Dineidin or Dineidyn. Din being the Brittonic equivalent of Gaelic *dun* meaning hill or hill fort and found in old Welsh as *dinas*. We don't get Edinburgh, spelled in all manner of different ways, until around the 10th century.

Prof. A.   Which is roughly when the Gaelic language influence was growing in the Lothians. Although the Gaels, have kept the old name, Dùn Èideann.

Dr J.   In which Èideann has no meaning. Just a name.

Prof. A.   Just a name. Right. Well, getting back to names of people, we have Cynon Eiddinn, son of Clydno Eiddin.

Dr J.   Remind me.

Prof. A.   According to Y Gododdin, you know, the poem about the band of warriors who went down from Edinburgh Castle Rock to do battle against the invading Angles, and who were hopelessly outnumbered and who were annihilated. Cynon Eidinn was the only survivor.

Dr J.   So Edinburgh was named after the only survivor and the name was applied to the fortress after the battle?

Prof. A.   Well, possibly. Or it was named after his father, Clydno.

Dr J.   What year was the battle?

Prof. A.   Oh, I'm bad at dates. Around AD600.

Dr J.   Do we have anything earlier?

Prof. A.   Yes. There's Áedán mac Gabráin.

Dr J.   Remind me.

Prof. A.   King of Dál Riata from roughly AD574 to 609. His kingdom was in western

Scotland, but there are records describing his sorties to other parts, including the east coast, and he certainly meets the criterion of a dominant person. For a while he was the most powerful person in Scotland.

Dr J.   And anyone earlier?

Prof. A.   There is. We can go back to the Romans.

Dr J.   No Roman name for Edinburgh?

Prof. A   No, but there is a record of a man called Cunedda ap Edern. The Romans, as I don't have to tell you, called the tribe that lived around Edinburgh the Votadini. And Cunedda ap Edern is said to have led Votadini forces against the Picts invading from the north. The Romans would have supported this. Anything to stop the annoying Picts up on their northern frontier.

Dr J.   What's the date?

Prof. A.   Around the end of the 4th century AD.

Dr J.   Hmm. Certainly meets the criterion of a dominant person.

Prof. A.   Right. And there's also his father. We have his name as Edern or Edeyrn.

Dr J.   Anything earlier?

Prof. A.   Not that I know of.

Dr J.   I suppose there's always Odin.

Prof. A. Mmm. So much for personal names. What about the other approach, natural features of the landscape? There's nothing down that road, is there?

Dr J. Not so sure. I might have one.

Prof. A. Something new?

Dr J. Well, look at the shape of the Castle Rock. And we're going back at least 1600 years, maybe 2000 or more. So there are no buildings round about. And very little on top. And north and south was boggy ground, perhaps even lochs, certainly a lot of water.

Prof. A. So what does it all look like?

Dr J. A fish's fin. For which a Welsh word is *adain*. Spelling variants *edn* and *edyn*.

Prof. A. Interesting. But you're just clutching at homophones. It's all in the realm of conjecture. As is everything else we've been saying. There's no getting away from, if you want a definite, clear-cut answer you're up against a brick wall.

Dr J. Or a rockface.

Prof. A. Ha ha. Your round.

Dr J. Aye.

Fairmilehead

This is another of these names that don't really stand up to too much inspection before cracking under the spotlight. A possible origin lies in two Gaelic words *faire* [faruh] (lookout) and *meall* [mel] (mountain). Say all that with a local accent and you should be getting pretty close to Fairmile. The –head is simply English tacked on as an afterthought.

Figgate

This odd-looking fellow is found these days only in a handful of names such as Figgate Park and the Figgate Burn, which latter is the name that the Braid Burn takes on as it gets closer to the sea by Portobello. Figgate is Nordic in origin, *gata* being street and *fé* being cattle. So really this is a Nordic version of the Cowgate.

Fountainbridge

Do you see a pretty fountain splashing and sparkling by a quaint and rustic medieval bridge? Scratch the surface and the picture changes. Earlier references to this show the name more than slightly different: as Foulbriggs or Foulbriggis. There was a stream here called the Foul Burn where the stinking sewage from the city of Edinburgh flowed, or rather where it clogged the burn. Later burghers, in a spirit of spin, and with more of a focus on the affluent than the effluent, saw to the sanitizing name change, as the area itself became less of a bog.

Gayfield

The Gay– here does not correspond to any of the modern uses of the word. Rather it is used in an older Scots sense of 'beautiful, pleasing to the eye'.

Gilmerton

The –ton is clear enough and standard enough: from Old English *tún*, which is a farm or settlement. The Gilmer– comes down from the Gaelic *Gille Moire*, which was a personal name with the meaning of son (or servant) of Mary. Its modern version is Gilmore. At the end of the 12th century the name of this district was recorded as Gillemuristona. The mixture of Old English and Gaelic stems from the period of some one hundred and fifty years or so from, in rounded figures, 950 to 1100 when there was a Gaelic-speaking, land-owning aristocracy in the Lothians, a minority in a region where most people spoke an Anglian language.

Gogar

There is a degree of mystery about this odd name, but one appealing theory is that it comes from Cumbric *cog* (cuckoo) with the ending *ar* for an arable field. Scots *gowk* is a cuckoo too. Cuckoo field is not unknown as a placename in other parts of Europe. The cuckoo, a bird far more often heard than seen, is associated with several old superstitions and rituals. But a good deal of this area used to be nothing but marshland, land to be avoided, hardly arable, not ideal for cuckoos and hardly likely to provide a significant enough characteristic to generate a name.

Name variants of Gogar down the years have included Coger (1336) and Grogar (1542). It is helpful to look for similar namings elsewhere. Are there other places with similar names and similar natural characteristics? There is a River Cocker in Cumbria. And another in Lancashire. East Coker in Somerset has a name recorded in the Domesday Book (1086) as Cocre, defined as a Celtic name meaning winding or crooked. So a more plausible theory is that the name Gogar (once Coger) is ancient British. There is also a connection with Welsh *crwca* (crooked), Gogar also once having been Grogar.

As so often happens, it is more than likely that the name Gogar was first applied to a watercourse, to the Gogar Burn, which wound its way through the boggy land and which would certainly have been a major characteristic, a significant enough feature to become the source of a name.

GORGIE

The earliest recorded forms are 12th century Gorgin(e) and 13th century Gorgy(n)(e). An old Cumbric word *gor* means big. The –gie element is less clear-cut, although there are two possibilities: a Cumbric word for field or enclosure (Welsh *cae*) or a Cumbric word for wedge (*cyn*) (which could be either a shape of the land or an implement for splitting stones or slate). In general it has to be said that fields are responsible for more placenames than wedges.

GRANTON

Here are two existing naming candidates. One says the name is from Old English *grene dun* which means green hill, describing the land which rises up here from the southern shore of the Firth of Forth and was the site of the now demolished Granton Castle. The other develops a form of the name that was recorded in 1505: Graintone. And relates this to Old English *greósn tún* which, with *greósn* being gravel or sand, is something like the farm or homestead by the gravelly shore. However, a third possibility is to be found in an old Scots word *grand*, defined as 'a spit of rocky beach or low-lying land running out into the sea'. Which matches Granton to a T.

GRASSMARKET

From the late 15th century onwards this picturesque place was home to one of Edinburgh's biggest markets. Horses and cattle, which had come up along the Cowgate, were taken to market here, but the place itself took on the name of their fodder which was stacked up here too.

Edinburgh's Grassmarket, pictured here in the early 18th century, was a centre for buying and selling horses and cattle. From 1660 for just over a hundred years it was also Edinburgh's public execution site, of which more than a few were carried out. The structure to the right is called the West Bow Well and is part of a series of wells that supplied water to this area and on down the Royal Mile from a reservoir up on Castle Hill.

*Overall the appearance of the Grassmarket has not
changed that much since the old days. The West Bow
Well is still there, minus its pump handle. But there
hasn't been a cattle or horse market here since the early
20th century. Activities are shopping, dining, drinking
and taking in the atmosphere.*

Gyle

Gaelic *goil* basically means boil or boiling but is also used to describe swirling water, which can give the appearance of boiling. The area was notorious for flooding; it was marshland by the old Gogar Loch, which was drained in or around 1766.

Hailes: Easter/Wester Hailes

Old English *heall* is a hall or a grand stone residence and is a word that is the likely origin of Edinburgh's two Hailes.

Hermiston Gait

A record dated 1233 contains the non-condensed Old English version of this name: *Hirdmannistoun* which is herdsman's homestead. The Gait is Old English *geata* which means either road or pass.

Holyrood

Rood is an Old English and old Scots word for cross, the cross of the Crucifixion. King David I founded the abbey here in 1128 and the name

Holyrood Palace and Abbey
© National Library of Scotland

Holyrood refers to the fragment of the 'true cross of the crucifixion' that, it was said, his mother Margaret had brought to Scotland. These days the name also refers to the Scottish Parliament just across the road.

INCHMICKERY

The Inch in placenames usually goes back to the Gaelic *innis* which means island and is pronounced eensh. And that is part of the usual explanation of this name, with the rest being Gaelic *nam bhiocaire* [nuhm vikuhruh] (of the vicar). But Gaelic name origins are not so common in this part of Scotland; related Cumbric offers another possibility. Cumbric/Welsh *ynys* is a related contender to account for this origin, also meaning island, and together with old Welsh *micariaid* this gives vicars' island.

INVERLEITH

In early texts Inverleith includes the district now known as Leith. But the reference of the name has nowadays shifted upstream a little so that this combination of Gaelic *inbhir* (inver) meaning mouth and Cumbric *llif* meaning river is more than slightly misplaced.

JOPPA

The name of the district is from the name of a farm that was once here and that was called Joppa after the name of the biblical city (now Jaffa).

KAIMES

The name is likely to stem from the crags on which an Iron-Age fort once stood and which would have been described in Old English as *camb* (and in related Old Norse as *kambr*), both meaning crest. The name was recorded in 1490 in the plural as *camys*.

LASSWADE

A name that, unlike many others, has not changed over the centuries; it comes from Old English *læss* (meadow) and *wæd* (ford).

LAURISTON

This was the *tún* or farmstead of Laurence (first available record 1290), a man, very likely of Norman descent, whose name has been slightly trimmed down by time and repetition.

LAWNMARKET

There is no connection with the Grassmarket, at least not insofar as the names go. The Lawnmarket is now one of the many streets, with different names, that link together to form the Royal Mile. The Lawnmarket's charter, when it was first drawn up in the late 15th century, described the place as a market for the sale of merchandise from the 'land', the distinction here being made with merchandise from the burgh. Goods that were imported for sale from outside the burgh were liable to tax, whereas goods that were sold by merchants resident in the burgh were exempt from tax. It was mainly cloth that was on sale here, linen in particular. The name, originally based on this charter's description, grew up as the Land Market; but with the passage of time it became known, in its written form, as something which, at first sight, has no connection with its function or purpose. But the word *laun* both with and without a final d is a Scots word for land. So, if this is a corrupted form, it is corrupt solely in the anglicization of the spelling.

LEITH

A connection is suggested with Cumbric *lleith* (modern Welsh *llaith*) which means wet or damp; and it is also likely that this naming element is the same as that which occurs in the second syllable of Linlithgow (the pool (lin) in the damp hollow (gow)). But it is unusual for a placename to originate just from a solitary adjective; and there is no evidence that *lleith* also meant damp place, although dampness and bogginess are at the root of many a placename.

But there are other reasonable alternatives. One is the Old English *hlith* meaning slope. And another is more Cumbric in the word *llif* which can have the sense of river. This is a stronger contender, river here outweighing slope as a likely reason for giving a place a name. And the shift from th to f is a well-known occurrence amongst speakers of the languages of Britain. Whichever its origin, as the language spoken here changed, the name lost any meaning and had reference but not sense. The Water of Leith, as the river that enters the sea here is nowadays known, means, on the analysis of the foregoing, no more than the Water of River.

LENNIE

This district name is very probably of Gaelic origin. A Gaelic *leanaidh* [lee-ahnee] is wet riverside land or a waterside meadow and is a word which also underlies the name of the town of Lenzie, just outside Glasgow.

LIBERTON

This was the *tún* or farm, specifically the *bere-tún* or barley farm, that was on a slope or *hlith*. The elements of the name are all Old English.

LITTLE FRANCE

These days there is nothing particularly French about this area. But clearly there used to be. In 1561 Mary Queen of Scots, as the recently widowed 19-year-old Queen Consort of France, came back to Scotland, landing in the harbour at Leith, which English forces had just the year before retrieved from French occupation. She took up residence in nearby Craigmillar Castle. It is possible that the royal entourage of French courtiers made quite an impact on the locals here. But it is equally possible that there was a longer-standing connection with an immigrant community from across the Channel, the city being traditionally cosmopolitan.

LOCHEND

In the early 12th century lands here were granted to the Norman de Lestalric family by King David I. The de Lestalrics then set about building a castle on the land. This was Restalrig Castle, later to be known as Lochend Castle and nowadays as Lochend House. The area around about the old castle was known as Lochend. This is shown in a map from around 1682 and is certainly an older name. So this was the castle and the district at the end or head of the loch. But the strange thing is that the loch, which should really account for this district's name, is itself called Lochend. Perhaps the name Lochend could be a corruption of the word lochan (little loch). Or more likely the loch itself, being so little, never actually had a generally accepted name and was itself named retrospectively after the district to which it had unwittingly given a name.

LOCHRIN

The name was not the name of a loch, but it does have connections with a loch, or rather with a burn near a loch: a burn called the Lochrin Burn. The loch in question used to exist on Edinburgh Meadows, was known as the Burgh Loch or South Loch, and supplied some of Edinburgh's drinking water up until the early 17th century. The old Burgh Loch drained into the Lochrin Burn. A 'loch run' is the name for such a drainage channel. And *rin* is an old Scots word for run. Hence Lochrin.

In the early 1600s the Burgh Loch, seen here, occupied much of what is now The Meadows. An outflow from a loch was known as a loch-rin, providing a name for the nearby little district of Lochrin. The clock tower of George Heriot's School (foundation stone laid in 1628) can be seen on the north side of the loch with the crowned spire of St Giles' Cathedral on the right.

The draining of the Burgh Loch and the passage of time and the growth of trees have created the green expanse of The Meadows. Even if the trees were not there, modern buildings now hide George Heriot's and St Giles from view.

LOTHIAN

This is thought to derive from a Cumbric personal
name, Leudonus. Leudonus (6th century AD)
is said to be a long-ago king of Leudonia, an
ancient name for this part of Scotland. And the
name Louthion, referring to these same parts,
is recorded many centuries later. According to
legend, Leudonus himself was the grandfather of St
Mungo, the patron saint of Glasgow.

MARCHMONT

Marchmont was developed by Sir George
Warrender in the mid 19th century and he chose
the name for the new suburb which he was building
on account of a family connection on his wife's
side. She was the daughter of Sir Hugh Purves-
Hume-Campbell, 7th Baronet of Marchmont
House in the Scottish Borders. As to the name
itself, there is one theory that it comes from an old
Gaelic word *marc* (horse) and *monadh* (hill). But a
reference to a hill in the borderlands or marches is
also a prime contender.

MERCHISTON

This is another intertwining of two language
threads. The –ton is Old English, the standard
word for farm or homestead. The first element of
this placename comes from a personal name, from
Merchion, and appears in a 13th century record
as Merchinston. Merchion is an Old Welsh name
and Merchion himself was doubtless a Cumbric-

speaker. This derivation casts light on the modern pronunciation of Merchiston, in which the ch is not like the ch in church but a hard k sound.

MOREDUN

The Gaelic is *mòr* (big) and *dùn* (fort) and the name clearly points to an early medieval fortification. But an ancient fort here? There's no trace of one. The name starts to make sense when you realize that this name only came to be used when the estate was bought by a certain Baron Moncrieffe in 1769 and that the Baron renamed the house and estate Moredun after the ancient hill fort in his own home county of Perthshire, the hill on which the hill fort stood also being known as Moncrieffe Hill. This is vanity naming or nostalgia naming and a reminder that expecting a necessary connection between placename and place without the mediation of human whim can be a trap.

MORNINGSIDE

This name was most probably given to the area because of its south-facing aspect. Morningside would get good morning sun.

NEWBATTLE

The –battle does not refer to a battle but is from Old English *botl* which means simply house. In 1735 it appeared on a map as Newbottle, which was probably regarded as just too strange.

NIDDRIE

Niddrie's name stems, with some of the edges worn off with time, from Cumbric *newydd* (new) and *tref* (house, homestead). A modern Welsh-speaker would immediately recognize the Cumbric name.

OXGANGS

This was once land under the plough. An *oxgang* is an old Scots (and English) term for an area of land, specifically the area of land that you could reckon on one single ox being able to plough over the course of one ploughing season. The exact area would vary according to the type of land or soil, but generally speaking would be around 13-15 acres.

PEFFERMILL

Two burns run through the playing fields here: one is the Pow Burn (this Pow probably linked to Old English *pol* meaning pool or slow running water) which joins the other, the Braid Burn. It is thought that one of these two once went by the name of the Pefr Burn, which in turn gave its name to a mill. *Pefr* is Cumbric (and now Welsh) for radiant, clear and pure and was clearly a significant feature of the flowing water here. This history is retained in the names of nearby Clearburn Road, Clearburn Gardens and Clearburn Crescent. There is another Peffer Burn near Aberlady, just east of Edinburgh.

PENICUIK

From Cumbric *pen* (head, top), *y* (the) and *cog* (cuckoo). So: cuckoo hill.

PILRIG

Two possibilities. One of the meanings of the Scots word *rigg* is a long, narrow strip of land that is ploughed and farmed and that is surrounded by rough uncultivated land. Tenant farmers would each have their own rigg. It is likely that a rigg here was closed off by a stockade or fence, called a *peel*. Another, quite different, possibility is that the name comes from Old English *pyll hrycg*. A *pyll* is a little creek, a channel down which water off marshy land flows into a river. And *hrycg* is ridge.

PILTON

In Scots a *peel* can be anything from an area surrounded by a stockade to a stone tower (known as a peel tower). In placenames the word *peel* has often been trimmed down to *pil*, making of Pilton the *tún* or farm or homestead with a *peel*. But, just as with Pilrig, there is another possibility: the *tún* or farm by the (Old English) *pyll*, a *pyll* being a creek feeding a stream.

PLEASANCE: THE PLEASANCE

A convent dedicated to St Mary of Placentia once stood here. In spite of appearances the saint's name has no connection with our placenta. In modern Italian the name has become Piacenza and it all comes from the Latin *placere* (to please). So The Pleasance is a pleasant place to be.

POLWARTH

From Old English *worth* (field) coming after the personal name Pol.

PORTOBELLO

Well, there is a good sandy beach here but no fine harbour as the name claims. And Portobello is not this Edinburgh suburb's first name: it was originally known as the Figgate Whins. The modern name, it is said, originates with the whim of a retired sailor, a certain George Hamilton, who had been at sea with Admiral Vernon and who had, in 1739, been present at the capture of Porto Bello in Panama from the Spanish. Coming back to Scotland in 1742, he named his cottage Portobello Hut in memory of his adventures. The hut's gone, but the name spread.

RATHO

The name, with spelling variations, goes back to the 13th century. A probable origin is Gaelic *ràth* [ra] which means a fort, often a circular fort. And *ràthach* [ra-oK] describes a place which has a circular fort. Scottish pronunciation superseded the Gaelic. There is no trace of a fort in Ratho but both Kaimes and Dalmahoy Iron-Age hill forts are nearby, which would have given good reason for naming.

RAVELSTON

This was once the *tún* or farm or homestead of a certain Hrafnkell, a man of Norse origin, or maybe it is safer to say, a man who had a Norse name. Hrafnkell is a name that occurs quite often in the Domesday Book (which of course did not cover land ownership in Scotland) but that was clearly a name in need of some flattening out for local tongues.

RESTALRIG

[pronounced: wrestle-rig]

In the 12th century the lands here were recorded as Lastalrik and belonged to Peter de Lestalrik. The form of the name starting with an L persisted into the early 16th century, when the form starting with an R began to take over. It is likely that the land was given to Peter de Lestalrik in return for knight service, but quite possibly the original land granted was in the North of England, with the Scottish land being a follow-up. But how to account for the name of Restalrig or Lestalrik? There are two explanations. One is that it may come from a dialect word *lestal* meaning mire or bog. The dialect is North English or Anglian, not Scots. And the –rig can be ascribed to either Old English *hrycg* meaning ridge or to Scots *rigg*, meaning a strip of land. The second possibility concerns the word *laystall*. In the Middle Ages a *laystall* was a stopping-off place for cattle being driven to market and was often – not surprisingly – associated with the secondary sense of the heaps of cow dung that would build up in such a laystall. The –rig again being the strip of land where the laystall was. Whether or not either of these features of the landscape actually existed here outside old Edinburgh is probably not the point, since the name was an import, a name the sound of which (more probably than its sense) appealed to a family of Norman nobles, putting down roots in a new country.

Riccarton

This is just Richard's *tún* or settlement or farm, a name that has changed very little down the years. First recorded as Ricardestone in 1296 and then just 10 years later as Richardtoun. No doubt a Norman name.

Roslin

The first recorded form of the name, in the year 1183, was Roskelin. Modern Roslin (or Rosslyn, as the famous chapel has it) takes its name from the old Cumbric words *ros* (moor) and *celyn* (holly).

Saughton

[pronounced with the *augh* as in l*och*]

This was the *tún* or farm or settlement by the willows, the Saugh– coming from the Scots word written either *sauch* or *saugh* (willow). The area is well-known these days for its prison.

Sciennes

[pronounced: sheens]

In the early years of the 16th century a convent was built here and named after St Catherine of Sienna. French being the court language of the day, Sienna was pronounced Sienne. The convent itself was demolished during the Reformation in 1559, but the name remained and was gradually transferred to this district, showing early signs of mutation, occurring in subsequent 16th century records as Shenis. Even this variant proved too much for the locals who whittled it down to the monosyllabic *sheens* while at the same time the written form underwent further random and irrelevant alteration.

Silverknowes

A Scots *knowe* is an English hillock or knoll. Silvery grasses here on the hills rolling down to the Firth may account for the first part of the name.

This is an artist's impression of the convent dedicated to St Catherine of Sienna which stood here in the early 1500s until, after a fairly short life, it was demolished during the Reformation in 1559. Arthur's Seat looms in the background.

The only remaining trace of the 16th century convent is the name of the district, spoken now as sheens, which, together with St Catherine, also occurs in several local street names.

SLATEFORD

No mysteries here, it is just what it says. The original village grew up around a ford over the Water of Leith and close to slate quarries.

STENHOUSE

This is simply stone house, *sten* being a variant of the commoner Scots *stane*. Stenhouse Mansion is the oldest building in the area.

STOCKBRIDGE

This was the place with the timber bridge. Old English *stocc* and Scots *stock* (log, trunk) account simply for the first element. These days, of course, the bridge over the gully where the Water of Leith runs, is no longer wooden; and the one-time village is now a city district.

SWANSTON

Some birds held special significance for our ancestors and the names of these birds sometimes found their way into the names of places. Swans are not known to be part of this set. The Swan– in Swanston is most likely to have come down from a man called Svein, a Norseman or a man of Norse stock, who built and owned the *tún* or farm or homestead here.

TOLLCROSS

This seems like a nice straightforward name for a cross-roads where tolls were collected. Logical. Until, that is, you realize that the name itself goes back to (at least) the 15th century when there was neither a cross-roads here nor would it be likely that, out here, way out of town, there was any thought given to collecting tolls. The –cross is far more likely to be the same cross as occurs in Corstorphine, that is to say a slightly re-arranged old Cumbric word *cors*, which is also old Welsh, and which means bog, swap or fen. And *toll* in that same linguistic origin means full of holes. Originally, therefore, a pretty nasty place to get stuck in and certainly an area to skirt around.

TORPHIN

This is from Gaelic *tòrr* (hill) and *fionn* (white, fair). That the name is unlikely to be connected with the –torphine as it occurs in Corstorphine (from the Orkney earl Thorfinn the Mighty) is attested to by the existence of nearby Torduff, which is the Gaelic opposite of white hill – *tòrr dubh* meaning dark hill or black hill.

TRON

The tron in any Scottish town was a public weighing place.

Turnhouse

This is one of those names that seem sensible enough until you start to think harder about it. Old records going back to the early 17th century show various spellings: Turnoch, Turnehouth and Turnheath, as well as Turnhouse. It seems quite likely that the first part of this name comes from Old English *þorn* which means thorn (with the þ pronounced as the th in thing). The second part varies according to which older version is chosen, but that the name meant something like thorny height is likely.

Wardie

Where the high ground rises up overlooking the Firth of Forth is where the first Wardie was. The name has its roots in Old Norse *varði* (beacon, cairn) as well as in Old English *wearda* (watch), a *weardan hylle* being a beacon hill or a lookout point.

Warriston

A 1467 record of this name has it as Warenstone. And we know from the lists of knights that sailed to England with William the Conqueror 401 years before that de Warenne was a Norman name. Warristone was the *tún* or farm of a descendant of the Norman Warenne, or at least of a man who bore a name of Norman origin.

Looking down on 17th century Glasgow from the hill of the Necropolis: the Molendinar
Burn in the foreground, then cathedral manse gardens, the Drygate and the spires along the
High Street beyond
© National Library of Scotland

GLASGOW

Anderston

Unlike many other placenames ending in –ton the origins of this Glaswegian district are not traceable to any early medieval settlement or Old English *tún*. In fact long before the area acquired its present name it was known as the Bishop's Forest, the land having been settled on the Bishop of Glasgow in 1450 by King James II of Scotland. The switch to something closer to the modern name came about in the 18th century when the area was first known as Anderson Town, after the Anderson family who owned the land here and, in particular, after James Anderson who, in 1725, leased out land where workers' cottages could be built. The name became condensed to Anderston. Towards the close of the 18th century Anderston, still a separate burgh and not yet part of Glasgow, was booming, noisy and smelly. It was a place with weavers (who had moved on to working with the new-fangled steam-powered looms), dyers, glassmakers, brewers, printers and others. In the 19th century huge warehouses went up to hold the merchandise off the Clyde. But the industrial boom had its day and the late 20th century saw the demolition of the old heart of the community at Anderston Cross in order to make way for the Kingston Bridge and the M8 motorway which swept over it and across the Clyde. International hotels and office towers now dominate. And the name of Anderston is fast becoming associated mainly with the local railway station.

This is Anderston farmland as it was around 1760, well outside the city boundaries. In the background the steeple and tower of Glasgow Cathedral can be seen far left. The spire far right is the Briggait and, just left of it, the new St Andrew's Square Church. At the far end of the tree-lined avenue, then known as Anderston Walk and now Argyle Street, stands the Trongate.

Intervening Anderstons have been demolished and only the railway station, huddled under flyovers, now displays the name. The Clydeside Expressway meets Argyle Street here and off- and on-ramps lead to and from the Kingston Bridge, one of Europe's busiest road bridges, carrying traffic over the Clyde.

ANNIESLAND

Who's the Annie in Anniesland? Many have conjectured, nobody's sure. What do you find if you go back in time? The area here was originally open country and farmland and formed part of the Jordanhill Estate (up to 1809). There was also mining for coal and ironstone. From the fact that Crow Road, the main road passing right through Anniesland, is named not after any crow but takes it name from Gaelic *crodh* [kro], which means cattle and which was so called on account of this having been the drove road along which cattle were driven to market, it is reasonable to conclude that the name Anniesland too might itself have a Gaelic origin. Maybe there never was any Annie. Here are two possibilities.

Maybe there's a relationship, a similarity in naming origin with a little district called Annathill over to the northeast of the city. There's an old Gaelic word for church, especially a church with a patron saint or a church where a saint's or holy man's relics were kept. The Gaelic is *annaid* and the (modern) pronunciation is ann-etch. But this would take the name far back into the Middle Ages.

A second theory is this: Gaelic *ainnis* means poor and destitute and the word *lann* can mean home or building. If Anniesland was originally no more than a home for the destitute, that would explain why it was slower to appear on maps as the name of a district (only Jordanhill and nearby Temple feature on a 1654 map). It is also perhaps not a coincidence that the learned minister of the parish, the Reverend Duncan Macfarlan, writing his 1836

Statistical Account (a very detailed set of parish records) refers to the district as Annisland, with no e. And if the –land was never –lann, it might well have started out as –land. The merging of Gaelic-derived terms with an English term is by no means unknown in the development of placenames (compare, for example, Annathill).

Auchenshuggle

A lot of people from Glasgow will know of this place simply by virtue of the almost legendary familiarity of the name. Many will be hard put to it to tell you just where Auchenshuggle actually is. And there may even be some who think it is a made-up jokey name, an old comedian's creation. Auchenshuggle was once a field and hung on to its resonant name, which comes from the Gaelic *achadh* [ahK-uhG] (field) and *an t-seagail* [uhn cheguhl] (of rye).

Auchinairn

The first part is certain. *Auchin* or *Auchen* in Scottish placenames refers to a field and comes from Gaelic. The second part is not so clear. The modern Gaelic name is *Achadh an Fheàrna* [ahK-uhG uhn yarnuh] which means field of the alder. But there are other possibles. It may be a contraction of Gaelic *achadh an eòrna* [... yornuh] which is barley field. There is also a suggestion that it might be from Gaelic *achadh na h-Èireann* [... nairn] which means field of Ireland and which gets closer to the modern pronunciation, and is backed up by a Timothy Pont map of 1596 which has it as Achinerin.

BALORNOCK

Here's a name which, at first blush, looks as though it should be built up, like so many others, from the Gaelic *baile* (homestead, farm). However, in 1186 the name was recorded as Buthlornoc and just 13 years before that as Budlornac. This first name element is thought to be from old Cumbric *bod* (dwelling, home), related to an older Gaelic word *both* (hut, cottage – giving modern bothie). The –lornoc is less clear but is most likely to be a personal name.

BARLANARK

From old Cumbric *barr* (although this was a Gaelic word too), meaning height or top. The second part is old Cumbric *llanerc* (glade or clearing), the overall reference probably being to a clearing at the top of a hill or a bare hilltop.

BARLINNIE

Barlinnie in the district of Riddrie was, to start with, the name of a farm; and the farmland was bought up in 1879 by the local authorities in order to build a new prison on it. The Bar– is most probably from Gaelic *bàrr* which means upper part or top. And *linne* is Gaelic for pool. So the site and name of this Glaswegian prison, known locally as the Bar-L, stem from a tranquil pool up on high ground.

BARMULLOCH

The name certainly fits the elevation of this suburb from which a good view of a large part of the city of Glasgow can be had. A *mullach* in Gaelic is a top part and the Bar– is also height or top. So this might be Top-of-the-hill.

BATTLEFIELD

Unlike the word battle in many other Scots placenames where it means house and derives from the Old English *botl* (house), this battle was a real battle. In 1568 the Battle of Langside was fought here and the forces of Mary Queen of Scots were defeated by the troops of James VI. The battle lasted some 45 minutes.

BEARSDEN

There are several theories about this name; here are two. First, it's not bears but boars. The name comes from Old English *bár* (boar) and *denu* (valley). Secondly, the name is old Scots, coming from *bear*, a variant of *bere* (barley) and *den* (small glen). In any event, the name only gained currency in the mid 19th century when it was transferred initially to the then new railway station from the name of a nearby house (and small district), before being put to use for the whole of the district that grew up here.

BELLAHOUSTON

There is a place called Houston in Scotland and it is explained as Hugh's town or Old English *tún* (farmstead, settlement). But Bellahouston has no connection with any Hugh. The –houston is traceable back to the Gaelic *ceusadan* which means crucifix. And the Bella– here is not, disappointingly, Italian for beautiful but a corruption of Gaelic *baile* [pronounced: baluh] which can mean settlement or farm. The full Gaelic form would have been *baile na cheusadain* [baluh nuh Kaysuhden] or maybe *baile na ceusta* [baluh nuh kaystuh], which, if you say it over and over down the centuries, and an anglicized pronunciation takes over from the forgotten Gaelic, will become Bellahouston, the farm of the crucifix. But why crucifix? All that can be said is that this fits a general scheme of religious naming, quite possibly initiated by early medieval monks, but imitated by farmers (for instance Joppa in Edinburgh).

In 1871 this parkland, which was then part of the Polmadie estate, was bequeathed to the city of Glasgow in the will of Moses Steven, who had bought the Polmadie estate. His bequest included the colossal sum of £0.5 million 'for the relief of poverty and disease'. At that point the name had already settled down as Bellahouston, the original Gaelic name having been establised many centuries before.

BISHOPBRIGGS

The *riggs* or land for ploughing that once belonged to the Bishop of Glasgow, when this was a tiny hamlet way out of the city on the road to Stirling.

The second b in the name has just wormed its way in (by a quite natural process). Although an 1856 Ordnance Survey map shows this as Bishop Bridge, there is no known connection with bridges.

BLOCHAIRN

The Gaelic *baile a' chàirn* meaning farm by the cairn or farm with the cairn is the district's current Gaelic name. But is it also its origin? For one thing it is unusual for *baile* to become bl. Also the name Blochairn is not unique to this district: there is also a High Blochairn Farm not too far away, near Milngavie. The repetition is interesting in that it could point to the name as describing a type. Another explanation for the name of Blochairn is that it is Gaelic *blàr* (field) with something like *a' chàirn*. It was common farming practice to clear a field of rocks so as to create better space for agriculture. The rocks would be piled up in something like a cairn, called a clearance cairn, which may have served a dual or multiple purpose but which was basically created in order to clear the ground. Cairnfield is a known name.

BRAIDFAULD

This is straight Scots, the district's name coming from a farm that used to be here. *Braid* is broad. And *fauld* is the part of the outer field where cattle were put and kept in folds or pens so that the ground got a good dunging.

BROOMIELAW

Law is a Scots word for hill, a rounded hill especially, and occurs frequently in Scottish placenames. But how can this fit? There's certainly no broom left these days and it is also rather difficult to see this riverside thoroughfare and business district and rush-hour snarl-up zone as having anything like a hill. A *law*, however, is also defined as 'a mound of earth and shingle on the bank of a river' and as a place where nets can be pulled up out of the water onto the bank. The name of this riverside stretch of Glasgow comes down from the Broomielaw Croft. In the 1700s this was still a broom-covered riverbank, with a croft, some distance outside the main town itself, although the industrialization of neighbouring Anderston was steadily changing things.

CALTON

Known locally with its honorific definite article as The Calton this is, or was, a place of hazels, the name coming either from the Gaelic *calltainn* (hazel) or perhaps from the Gaelic *coll* (also hazel) with the English ending –ton from Old English *tún* (farmstead, settlement).

CAMBUSLANG

An ancient place with remains dating from the Iron Age, the name of Cambuslang is from Gaelic *camas* [kam-uhs] (bay) and *long* (of ships), with the *long* having undergone a quite natural Scottish mutation. The Clyde was of greater importance here in olden days, Cambuslang lying close to the end of the river's tidal reach.

CAMLACHIE

Camlachie's just about gone now. Though there's a street name left. The Forge Retail Park occupies some of its space. The name of what was once a village, then a centre for many industries, and for coal mining, comes from Gaelic *cam* (bend) and *làthachach* [pronounced: laKuhK] (muddy). The muddy bend was in the Camlachie Burn, which once powered mills here and which is now mostly underground.

CANDLERIGGS

Riggs are strips of land, often farmland, but here the word describes the place where candlemakers produced their goods and traded, the name eventually to become a street name in the modernized Merchant City of Glasgow. A row of candlemakers might seem unusual; there was a reason for it. In the summer of 1652 there had been a Great Fire of Glasgow. One third of the city burnt down. Some years later, in 1658, the authorities passed an act banning dangerous trades from the city centre. And this meant that four candlemakers had to relocate their businesses here, giving a name to this street, which was in those days just outside the city limits.

CARDONALD

This was the *caer* of Donald, a *caer* being a
Cumbric word for a fort or castle, although this
would not necessarily have been built of stone. The
Stuarts of Cardonald were the family who owned
the land here in the 15th century, being related by
marriage to the Stuarts that owned the land and
house at Castlemilk. The original Donald is not
known; and the *caer* is no longer there.

CARDOWAN

The Car– is most probably from Cumbric *caer,*
being a fortress or fortified enclosure of some
description and material. And the –dowan is a way
of writing the Cumbric *duon* which means black or
dark.

CARMUNNOCK

Carmunnock has two elements to its name. The
Car– is most likely not, what it frequently is,
Cumbric *caer* (fortress, castle) but Gaelic *cathair*
[ka-hir] in the sense of seat or home. And the
–munnock comes down from Gaelic *manach*
(monk). The present-day Gaelic name is doubtless
here also the origin: *cathair mhanach* (monks'
place).

CARMYLE

A common theory is that the name is Gaelic
in origin, deriving from *càrn maol* (bare hill)
or *cathair maol* [ka-hir murl] (bare seat). The

bareness was manmade, the once wooded riverside land having been stripped of trees and bushes so that the soil could be used for agriculture.

CARNTYNE

This is likely to come from Gaelic *càrn* (hill, cairn) and *teine* [pronounced: chaynuh] meaning fire. But Cumbric words do the job too with *carn* and *tan* (fire cairn). When references to fire occur in placenames this tends to mean that the place was used, at some time in the past, as a site for a signal beacon.

CARNWADRIC

This is a complex looking name. Carnwadric was originally an outlying farm. It is likely that the element for farm is contained in the ending, in the Old English *wic*, commonly found in Scottish names, with the w having gone missing. It is not unprecedented for the w in *wic* to be lost in the spoken form. Take, for example, the pronunciation of Berwick [berik] or nearby Fenwick [fenik]. Or in an older written form of Prestwick (Prestik), in which *wic* the w has these days come back again. The Carn– at the front is most probably as in cairn, a pile of stones or rocks that functions as a marker. And the –wadr– (once written wather) is most likely the name, Walter, of the owner of the farm. This then is the cairn marking the site of, or entrance to, Walter's farm.

CASTLEMILK

Long ago there was a castle here, a fort, and for
centuries the place was known as Castleton. In
the second half of the 15th century ownership of
the castle and its estate passed to the Stuarts of
Castlemilk from Dumfriesshire, who later renamed
the Castleton estate Castlemilk, aligning it with
their own heritage. The Stuarts developed the old
fort into a fine country mansion, which, however,
over the years gradually fell into decay. The local
council in Glasgow finally demolished the former
mansion in 1969, housing estates being the then
order of the day. But what about the −milk? The
−milk in the name comes ultimately from the name
of a Dumfriesshire river, the Water of Milk, which
flows where these Stuarts once lived. And it is quite
probable that the Gaelic word *mealg* meaning milt
or fish roe (or its genitive form *meilg* meaning 'of
milt') is in at the start of the naming process, with
the Castle− being either straight English or the
Gaelic *caisteal*.

CATHCART

The −cart comes from the name of the river here,
a river only about three-quarters of a mile long,
formed by the White Cart and the Black Cart and
flowing into the Clyde. And the Cath−? Older
recorded forms of the name are Kerkert (1158)
and then, very soon after that, Katkert (1170). The
earlier form of Kerkert points towards Cumbric
caer (fort); this requires an assumption that the
castle built in Cathcart in the mid 15th century

(of which the ruins have now been removed) was constructed on the foundations of a much earlier *caer*. A related Gaelic word *cathair* [pronounced: ka-hir] is another contender here. It means seat or home or sometimes fortified place; the th is not pronounced in the original *cathair* – but then Scotticizing the pronunciation of Gaelic words is a common fate of a name.

The family that occupied this land from the 12th century went by the name of Cathcart, so the origin of the name itself is better looked for in Katkert than Kerkert. As to the meaning there are two possibilities. It may be that the Kat– or Cath– is a form of a Cumbric word that survives today in Welsh. *Caeth* means confined, narrow. Or, perhaps more likely, the Cumbric *caer* (fortified place) and related Gaelic *cathair* were at some earlier date used interchangeably, some folk using one version, some another, until eventually it was the Cath– that won out. Right up to the early 19th century, as Cathcart parish minister, Dr James Smith comments, writing in his parish records in or around 1838: 'Carcart is still the vulgar pronunciation'.

As for the name of the Cart itself, that goes farther back in time. A link with an Old Irish Gaelic word *cart* (to clean, sweep away) has been put forward. And that is no doubt cognate with Cumbric/ Welsh *carthu* (to cleanse); and that is, of course, a function of flowing water.

CESSNOCK

A reference book published in Glasgow in 1805 and succinctly entitled: *The Scotch Itinerary, Containing the roads Through Scotland, on a new Plan. With Copious Observations, for the Instruction and Entertainment of Travellers* makes reference to a Cessnock Hall, then inhabited by one Henderson, Esq., just a few miles out of Glasgow town and in the Parish of Govan, quite close to Ibrox, as shown on a 1795 map. This Cessnock Hall must surely be the origin of the name of the modern district of Cessnock. This same 1805 publication also refers to Cessnock ruins by Galston, down in Ayrshire. The ruins would be Cessnock Castle. And also in that neighbourhood we find Cessnock Water, the river name presumably preceding the name of the castle. But the meaning, what is the meaning? *Cess* is a land tax – but that is not a likely connection. More likely is a Gaelic connection somewhat along the structural lines of not too distant Cumnock. The –nock in both is Gaelic *cnoc* [pronounced: nock], meaning hill, and the Cess– could well be from Gaelic *seisg* [shaysk] meaning sedge. Sedgehill is a known name elsewhere in the UK. The name in Glasgow looks like an import from Ayrshire and not a description of anything local.

Clyde, River

A traditional explanation is that the name of the river originates from Cumbric *cloid* which is claimed to mean the cleansing one.

But what do we know? We know that Tacitus wrote about the British campaigns of his father-in-law, the Roman general Agricola, around the year AD98. In Tacitus the Clyde is called *Clota*. That is the first recorded name for the river. We know that, up to the year AD870, the ancient Britons had their capital at Dumbarton Rock on the Clyde. In their time it was called Alt Clut or Alclut. We are informed that *al-* is a pre-Celtic, Indo-European word for a watercourse, being found in many British river names, such as Ale Water, Allander, Alne as well as others in continental Europe, such the Ala in Norway or the Aller in Germany. We also know that *clud* is an Old English or Anglo-Saxon word for a rock or a hill.

It is generally agreed that river names are among the oldest names that we have. Often a river name simply means no more than *river* – for example a river called Avon derives its name from Gaelic *abhainn* [ahvin] meaning just river. Names with the pre-Celtic element *al-* are much older than Gaelic names; but the root of these names just means flowing water. There was, in general, no need or reason for ancient Britons to differentiate one watercourse from another. But if such a need should arise, for example, if visitors from another civilization asked about the name of the river, then some major distinguishing feature would naturally be called on. And here what better distinguishing feature than the name of the great fortress rock that towered over the river, the Clut of Alclut? What Tacitus, or Tacitus' informant, heard in the course of a doubtless laboriously interpreted conversation between a group of ancient Britons and a map-making southern European explorer was recorded as the name *Clota*. The old British name first referred to the rock at Dumbarton, the river rock. But the Clut element would do very nicely to differentiate this particular river. And *Clota* or *Clut* became the Clyde.

Cowcaddens

Not cows but hazels, the Cow– stemming from a Gaelic word *cùil* [pronounced: kool] meaning corner or nook and placed in front of the Gaelic *calltainn* [pronounced: kaltin] (of hazel trees). The Scots pronunciation as koo-kaddens remains a little more faithful to the original than does the anglicized Cow–. An architectural mishmash and numerous sets of traffic lights have replaced the hazel trees.

Cowlairs

This is a part of the city that was once famous as the home of Cowlairs Locomotive, Carriage and Wagon Works, producer of steam engines and rolling stock exported to the world, a plant built in 1842 and taking its name from a nearby mansion, Cowlairs House in the Cowlairs Estate. Cowlairs

House had been a fine country retreat in the early 19th century, but was eventually abandoned as the railway lines and works cut up its land. When it was sold, the land here seems to have been referred to as Fountainwell, but there are magistrates' records from 1631 referring to the land at Kowlairs, so it can be assumed that the new owner re-employed an older name; and one which has stuck. Unlike other Cow– names this one does look like pure Scots: *kow* is cow and Scots *lair* is the place where an animal beds down as well as an enclosure for cattle.

CRAIGEND

This is from Gaelic *creag* (rock). These days the nearest equivalent to a rocky end is to be found in the two huge water towers that dominate the skyline.

CRANHILL

If the origin is Scots then the Cran– is a heron. If the origin is Gaelic and if Gaelic and English –hill have been put together then there is a possible connection with the Gaelic *crann* which means plough.

CROOKSTON

Neither villains nor old Scots hooks, this name was given to the one-time village by the landowner. He was an Anglo-Norman knight by the name of Sir Robert de Croc and in the year 1170 King David of Scotland made a gift to him of the lands hereabouts.

CROSSMYLOOF

[pronounced: cross-muh-loof]

There are three accounts for this strangest of names. One has it as stemming from the Gaelic *crois mo Liubha* [krosh mo loova] which means the cross of my Liubha, Liubha (the spelling is Gaelic) being said to be a saint or holy man or woman. Another popular theory is that Mary Queen of Scots passed by here in 1568 on her way to observe the fortunes of her troops at the Battle of Langside and that she was carrying a cross in her *loof*, a *loof* being an old Scots word for the palm of her hand (also written *luif*). The cross was doubtless a plea to providence. But there is also a story that a fortune-teller stepped out of the crowd here, as the Queen passed by, and asked her to 'cross my loof' (with silver) if she wanted to hear which way the battle would go.

DALMARNOCK

This name comes from Gaelic *dail* [dal], which means a field, and the name of a sainted priest, Iarnan. This is the same name as the one which occurs in the name Kilmarnock. The ending –ock is from an old diminutive –*oc*, expressing fondness. And, as often the case with placenames deriving from saints, the name is preceded by Gaelic *mo*, which means my. So this is literally the field of my dear little Iarnan.

DARNLEY

The Darnley Sycamore is said to be a tree under which, in January 1567, Mary Queen of Scots stopped to tend to her sick husband, Lord Darnley. Mary had come to nearby Crookston Castle, where Darnley had been staying, to take him back to Edinburgh. Whatever the historical truth of this traditional story, the modern district's name derives from its association with Lord Darnley (who was to be murdered shortly after his return to Edinburgh).

DENNISTOUN

This district takes its name from Alexander Dennistoun, the son of a much-respected and very wealthy Glasgow banker and cotton merchant, James Dennistoun. Alexander was born in 1790 and, at the age of 71, in the year 1861 he commissioned Glasgow architect James Salmon to design a new residential suburb which was to bear the name of Dennistoun. By the time of Alexander Dennistoun's death in 1874 only some of the grander dwellings had been put up.

DOWANHILL

It is possible that the Dowan– goes back to an old Cumbric word *duon* which is an inflected form of the word for black. That is not to say that this west end of Glasgow district has any ancient origins. The Dowan element has long since cast off from its origins and become a naming element in its own right but with no meaning. There is a Dowan Farm a few miles further northwest.

DRUMCHAPEL

Before it became a housing estate this was the ridge
of the horse, from Gaelic *druim* (ridge) and *chapuill*
(of the horse). *Druim* can also mean back, making an
alternative interpretation of horse's back a possibility.

EASTERHOUSE

Long, long ago, far out in the country to the east
of the old cathedral city of Glasgow, there was a
farm called Easterhouse. In the late 19th century
a village began to grow up on land that belonged
to the Easterhouse farm, out in the country. And
the village also took the name of Easterhouse.
And then in the mid 20th century this name
transferred to the spreading council estate that
was built here. There was once a Westerhouse too.
But Westerhouse only grew up to become a street
name. A street name in Easterhouse.

FINNIESTON

In or around 1768 a new village began to grow up
in this area when the landowner, Matthew Orr,
leased out parcels of his land. He named the new
development, in those days out in the countryside
and not part of Glasgow, in honour of his tutor, the
Rev. John Finnie.

GARNETHILL

The jewels on this steep-sided hill are architectural
and educational: Charles Rennie Mackintosh's
Glasgow School of Art and St Aloysius' College. No

gemstones have as yet been unearthed here. An 1857 map shows a street called Garnetbank at the foot of the hill, but has no mention of Garnethill, which was known earlier as Summerhill. The school up on the hill, St Aloysius, was founded in 1859 by Jesuits who arrived in Glasgow that same year. In the early 1860s they bought land in the Summerhill/Garnethill district, which, at the time, was a residential area with spaciously set out villas for the well-off on the western fringes of the city. But the name? The name was given in memory of Thomas Garnett (1766 – 1802), an Englishman from (then) Westmoreland who was appointed Professor of Natural Philosophy at Anderson's Institute in 1796 (this being the predecessor institution to the University of Strathclyde). But there is an implicit duality in the name and a happy coincidence which must have carried weight with, and pleased, the Jesuits, who would certainly have been aware of the nominal link with Saint Thomas Garnet (c.1575 – 1608), a Jesuit priest executed as a martyr in London two and a half centuries before.

GARNGAD

Garngad is the old name of Royston. It quite possibly stems from old Gaelic *gart na gaid* (field of the withe). Of the what? It has to do with a nearby stream, the Gad Burn. A *gad* in Gaelic can mean a withe, that is to say a flexible twig that could be used for carrying a catch of fish; the word in Gaelic was also used for the catch of fish itself. Why Garngad fell out of favour as a placename is not known.

GARSCADDEN

One theory has this as stemming from Gaelic *gart sgadain* [pronounced: garsht skaden]. But just a minute. This means herring field and does at first seem a little unlikely... But then 12th and 13th century charters do refer to the importance of herring fishing in the Clyde and it cannot be taken as unquestionable fact that the Clyde and its tides were always as they are now. For example, a map made circa 1600 shows Kingsinch, just across the water from Garscadden, as an island. Another possibility relates the name to the Glasgow district of Cowcaddens: so Gaelic *gart* (yard) and *calltainn* [pronounced: kaltin] (of hazel trees).

GARSCUBE

These are the squashed remains of old Gaelic *gart* (field) and *sguab* [pronounced: skoo-uhp] (of the sheaves).

GARTCOSH

From old Gaelic *gart* (field) and possibly *còs* which means a hollow or crevice.

GARTHAMLOCK

From old Gaelic *gart* (field, often used to mean a cornfield) with the –lock being an anglicization of loch. The main road that runs through the district is truer to the naming origins as Gartloch Road; it runs close to several lochs. The district's name comes from Garthamlock House, which, in the 1950s, the city council bought, demolished and replaced with a school. Long ago the land here was part of the estates belonging to the Bishops of Glasgow.

GARTNAVEL

From old Gaelic *gart* (field) with *na abhal* [nahvuhl] (of apples). Nowadays a big hospital, before that a lunatic asylum, as they called them then, and before that a farm.

GIFFNOCK

This is a Cumbric name coming from *cefn* (ridge) with the diminutive ending –*oc* (little).

GLASGOW

Unlike Edinburgh, the city of Glasgow need be in no doubt as to the origins of its name. The name is from Cumbric *glas* (green) and *cau* (a hollow). Gaelic *glas* means green or grey-green too; and the element *cau* meaning hollow has passed into modern Welsh. It is still easy to see this green hollow up where St Mungo built his church towards the close of the 5th century, between the modern cathedral and the Necropolis. The –*gow* element in the city's name can also be found across the country in Linlithgow. And the element *glas* is a part of many names too. Ben Glas is the grey-green mountain and Glassford's the green wood. But *glas* is a tricky word and not every occurrence is a Glasgwegian *glas*. *Glas* is also a Brittonic or Cumbric – as well as old Gaelic – word for water or stream.

Glasgow Cathedral as it was in the late 18th century in a setting that had not changed for many centuries. The cathedral sits on one side of the green hollow of the original Glasgow. Through the green hollow runs the Molendinar Burn. Part of the cathedral tower can just be seen far left behind the trees. This tower was demolished in 1846 – to great public outcry.

Looking at where the city started from the same spot today. The Molendinar Burn has been channelled underground. The hill rising up on the right of the green hollow is the Necropolis.

GORBALS

Unlike many or most placenames this one is always used together with the definite article. This is The Gorbals. The name has a curious history. In Scots legal language *decimae garbales* is the tithe (or, in Scots, *teind*) owing to the church, as the traditional tenth share of produce, payable for the support of the clergy. *Garbales* derives from the medieval Latin word *garba*, which means sheaf, as in a sheaf of corn. Centuries before development, clearance and redevelopment The Gorbals was, it seems, the source of tithes, of standing sheaves of corn in open fields.

GOVAN

[pronounced to rhyme with oven]

There are three alternatives. The first, though tempting, will be rejected: there is a connection with the Gaelic *gobhainn* (smith). Although the modern Gaelic name for this Glasgow district is *Baile a' Ghobhainn* (smith town) the *Baile* (town) is not present in the actual name, or in the historical records of it, and to call a district simply 'smith' rather than something like 'the smith's place' or 'place of the smithy' is improbable.

Alternative number two is Cumbric *cefn* (ridge). This is likely to match the pronunciation of Govan, insofar as we can reckon how Cumbric was spoken. But the only ridge here is the one visible across the Clyde at Partickhill, land that was once a royal hunting ground. It is true that the later Parish of Govan did extend both south and north of the Clyde, but the old settlement of Govan was south of the river. And names are not usually given on account of what can be seen *from* a place.

The third alternative is based on a Welsh word *cwfen*, which means monastery or religious community. This was a significant identifying feature of the place, more than adequate to be the source of its name. According to tradition, Kentigern, also known as St Mungo, established a monastery at Govan in the 5th century. And it is clear that Govan was a place of some significance, both religiously and politically. Evidence for the political aspect of this is provided in part by the Doomster Hill, an artificial raised and stepped mound sited a short distance from the old church. This was a moot hill, a place where those in authority would come to hear cases and dispense justice. The Doomster Hill was eventually removed to make way for shipyards. As for religious significance, archaeological investigation has discovered remains of a Christian cemetery dating back to the 5th century, pre-dating the Gaelic. And Govan was known as an important place, a stopping-off point, on the old route from Lindisfarne to Iona in the days when Scotland was being converted to the new religion of Christianity.

The city of Glasgow was to develop further up river, eventually reaching out to, and swallowing up, old Govan.

(Above) *The tranquil village of Govan at the confluence of the Clyde and Kelvin, as it looked in the late 1700s, hardly changed for centuries. The Doomster Hill can be seen on the left.*

(Top right) *Govan at the height of Clydeside ship-building in the mid-twentieth century. The ancient Doomster Hill was flattened to make way for the shipyards, but the church (arrowed) remains a constant, though here invisible, feature.*

(Bottom right) *Govan today (2018) with its parish church re-emerging, shipyards mostly gone, housing spreading. On the opposite bank of the Clyde, in the foreground, is the Riverside Museum.*

HAGHILL

In Scots a *hag* is a place where peat has been dug, leaving land that is difficult or dangerous to walk across. A hillside cut up and pitted in this way would certainly have merited a special name. A connection with hags, whether old women or witches, is less likely.

HUTCHESONTOWN

The Hutcheson brothers, George and Thomas, were Glasgow lawyers, landowners and philanthropists. George founded Hutcheson's Hospital (now defunct) in 1639 and Thomas founded Hutcheson's Grammar School (not defunct, going strong) in 1641. Over 150 years later, in 1794, the governors of Hutcheson's Hospital embarked on the development of a new suburb in hitherto unused and unusable boggy land. There was only one choice for the name.

HYNDLAND

This is the back land. So called because it was at the time of naming at the edge of town. Originally part of the Parish of Govan, as shown in a 1795 map of Govan as Hindland.

IBROX

This is said to be from Gaelic *àth* (ford) and *bruic* (of the badger), an explanation which fails, however, to account for the final –x sound. But then Old English *broccs* is also badger and with the passage of time, and the mixing and merging of peoples and tongues, original forms mutate. Another contender, seen through the mirror of Welsh, is the Cumbric language with *y brocc* – the badger.

INCHINNAN

From Gaelic *innis* [eensh] (island) and *Finnén*, a personal name, usually now St Finnan.

KELVIN, RIVER

The River Kelvin has lent its name to many districts of the city, Kelvinbridge, Kelvindale and Kelvinside. There are two possible accounts for the name Kelvin itself. It may be from the Gaelic *caol-abhainn* [kurl avin] which means narrow river or narrow part of a river. Or it may derive from a Gaelic compound *coille-abhainn* [koyl-yuh-avin] which means wooded river. (The banks of the river are, to this day, thick with trees and shrubbery.)

KINNING PARK

Kinning quite possibly comes from an old Scots word *cuning* which means rabbit. Motorways and industrial estates have taken over the parkland.

MARYHILL

A good many Scottish placenames containing the name Mary or the Gaelic equivalent *Moire* are formed with reference to the virgin Mary. But not this one. There was in fact a lady called Mary Hill and she lived from 1730 to 1809 and was the daughter of the Laird of Gairbraid. The laird owned the land here and, since he had no son, he left it to his daughter, whose name became the name of the district that grew up and replaced the estate. Gairbraid House, the heart of the old estate, was demolished in the 1920s but the name Gairbraid lives on in street names.

MILNGAVIE

[pronounced: mull-guy]

There are three contenders here, all arguing a Gaelic origin. Gaelic *maol na gaoithe* [murl nuh gur-yuh] (windy bare rounded hill) is one theory. *Muileann gaoithe* [moo-luhn gur-yuh] (windmill) another. And a third is *muileann Dhàibhidh* [moo-luhn GYvee] (Davie's mill).

The first theory doesn't fit the lie of the land here. The second would have us believe that a windmill was once built in a place where the Allander Water, racing through the centre, was seen as inadequate for powering a mill. The third totally mixes up the Gaelic pronunciation and what the name might look like to non-Gaelic speakers (a common occurrence with placenames). What has happened? The Gaelic *ui* has become an English *u* as in mull;

the Gaelic *n* has been kept in the English written form and dropped in the English spoken form; the Gaelic *dh* has turned (appropriately) into an English *g*; the Gaelic *–àibhidh* has been kept in the English written form (*–avie*) but trimmed back to a rendering as *–àibh* and then deprived of the last two consonants (bh) in the spoken English form.

Molendinar

An old word from medieval Latin *molendinarius* which means miller or, adjectivally, mill. The Molendinar Burn is the Mill Stream, the burn that powered the mills. It's not easy to find these days, rising in the Frankfield Loch then trickling into the Hogganfield Loch, up in the northeast of the city in lands that, in the 12th century were the domain of the bishops of Glasgow. It flowed down past St Mungo's church, now Glasgow Cathedral, and on to the Clyde, where near its mouth it deposited enough silt to allow the then bridgeless river to be forded.

Mosspark

Up to the end of the 19th century Mosspark was farmland. But mossy? In Old Scots (or Old English too) moss is not moss but boggy ground or marshland. In general moss was not necessarily useless boggy ground either, since the moss may have been a source of peat or, as here, land usable for farming.

Mount Florida

The name has changed, but only slightly. The earliest known written reference to this district is from 1814 when the Glasgow Herald printed a notice for the sale of the estate of Mount Floridon. A minister of the parish of Cathcart, Dr James Smith, writing in the early 1840s, referred in the church's Statistical Account to 'the ridge of the hill now known as Mount Floridon'. The name Floridon stands out as not having any of the characteristics of Scottish placenames. It is reasonable to conclude that it is either made up as a result of the original estate owner's whim and fancy or, looking to the only other known recorded use of the name, that it was influenced by the title of a 17th century French novel by Jean de Segrais.

Helping to muddle matters more, there is an 1819 reference to a sale on the same estate, calling it Mount Floradale. Perhaps some light can be thrown on all this by the Herald's reference to the fact that there were two lodgings on the Floridon estate. And an 1850s map does clearly show the house as a kind of semi-detached mansion, each with its own separate walled orchard. Did each half of the house have its own name?

In 1856 the Ordnance Survey name book records Mt. Florida and Mt. Floridon as 'variant modes of spelling the same name'. The Florida version was attested to by three individuals. However this doesn't look like a question of spelling.

Was there a connection with Florida? This has been claimed by Alexander Gartshore in a 1938 publication called Cathcart Memories, a book full of very detailed accounts and recollections, in which he writes 'For many years after Dr Smith came to the small country village of Cathcart (minister from 1828 until 1897), as it then was, the nearest building in the direction of Glasgow was Mount Florida House which stood on the hill where Eildon Villas now are, and which was occupied by a family from Florida, USA – hence the name Mount Florida'.

The house was destroyed by fire in or around 1855 and on a map of that decade it is described as a ruin, but a ruin by the name of Mount Florida. The modern Glasgow district took the name of the ruin, which, along with the estate, has now completely disappeared.

MOUNT VERNON

In 1742 the land here, known as Windy Edge, was bought up by a Glasgow merchant called Robert Boyd. He rechristened his new purchase Mount Vernon as a tribute to Admiral Vernon who had become a popular national hero on account of his naval victories against the Spanish in the Caribbean.

MUIREND

Muir is Scots for moor. This was the end of it.

NETHERLEE

The nether or lower of what were formerly called the Lands of Lee, lower meadow or lea.

NITSHILL

Scratchy kids? *Nits* is a Scots word for nuts. Here once they grew.

PARTICK

The old Cumbric or Brittonic word *perth* means bush or thicket, a word that remains in modern-day Welsh, and is indeed responsible for the city of Perth. The –ick ending, or something like it, was there way back in the 12th century when this place was recorded as Pertheck, in the 14th to become Perthik. Welsh *perthog* is bushy. Partick, the bushy place, and former royal hunting grounds.

POLLOK

The name is from Cumbric *poll* (pool) and the ending *–oc* (little). Gaelic is similar with *pollag* (little pool). It is thought that the land here was given to some incoming Norman knights in the 11th century in return for military services rendered by them, the knights thereafter giving themselves the title ... de Pollok.

POLLOKSHAWS

The *shaws* of the knights of the little pool. *Shaws* is a Scotticized variant of the Old English word *sceaga*, which means woods.

POLLOKSHIELDS

The shields of Pollok are the *shiels* of Pollok, a *shiel* being a Scots word for a shepherd's shack and a shelter for his sheep. Since the district's development began as late as the middle of the 19th century, if there were shiels here then, a clearance would soon have taken place.

POLMADIE

Importantly here the stress in the name is on the last syllable, on the –die [pronounced –dee]. It would seem that the Gaelic name reflects the original. *Poll Mac Dè* or *Poll Mhic Dhè* (in modern Gaelic) is the pool or stream of the son of God, although the pronunciation of each version has been well Scotticized. The district's name is likely to have originated with the name of the Polmadie Burn that flows into the Clyde here. Going by the name, we may conjecture a site for religious rites or maybe the name simply reflects a general predilection for using biblical references in placenames.

PORT DUNDAS

The name came into being around 1790 when the port on a branch of the Forth and Clyde Canal was named after one of the canal company's funders, Sir Lawrence Dundas. His own name (irrelevant here of course) is from Gaelic *dùn* (fortress) and *deas* (south).

POSSILPARK

This is nowadays known locally as Possil. The first mention of the name goes back to the 13th century when the lands or estate here were recorded as Possele and were part of the total gift of lands made to the Bishop of Glasgow in 1242 by the King of Scotland, Alexander II. Whether the land here was actually known as Possele is not known; all that can be said is that the land was recorded under this name. No obvious meaning attaches to the name. A little over 300 years after the gift of the land to the Bishop, the Possele was divided up into Nether Possil and Over Possil. It is quite likely that Possele was never a particular name at all but a general term, a variant of *parcel* or *parcelle* in the court French of the day, meaning no more than piece of land.

PROVANHALL

The district is named after the 15th century building called Provan Hall, the Provan being another word for the Preband or the Prebend, which is the land assigned by the church to a canon of the church as the source of income which the canon receives for exercising his duties.

QUEENSLIE

Once a farm and now an industrial estate this name has changed down the centuries more than somewhat. When a farm, in the mid 18th century, it was written Quinsley. *Quin* in Scots can be queen (as well as woman). But it could also be a sterile ram. The –ley or –lie is uncontentiously a meadow or field. Leaving the ram aside, this element with queen/quin might be thought to lend some weight to a traditional account of the origin of the name, which is that Mary Queen of Scots spent a night in the old steading of Easter Queenslie. But then a document dated 1667 refers to the 'lands of Easter Cunschlie and Wester Cunschlie' and it would be strange if, so soon after the Queen's supposed stay here, the name should have been so different. Perhaps the solution is pointed to by a document written in 1896 which refers to earlier names of Easter and Wester Cowhunchollie, the former being (in 1896) Cranhill and the latter Queenslie. Now Cowhunchollie is a Scots rendering of the better known personal name of Colquhoune. Queenslie was once Colquhoune's meadow.

RIDDRIE

There is a naming connection here with the village of Raddery up north on the Black Isle. Raddery has a Gaelic name Radharaidh, a name meaning arable land not in tillage, not being cultivated. Degaelicizing the pronunciation brings you to Riddrie.

ROBROYSTON

In the 1500s this part of the world was known as Roberstoun (simply Robert's farm or *tún*). Placenames often undergo contraction from a longer form to a shorter; the converse process of expansion from a shorter to a longer form is rarer. The name change here may well have been influenced by the emotive and romantic pull of the name of the Scottish folk hero Rob Roy MacGregor, who was born the following century in 1671. There is no evidence for this; it has to remain a reasonable enough assumption. The main historical association of this district is that it was here that William Wallace was captured and handed over to the English troops in 1305, probably before the place had any name at all.

ROYSTON

This district was originally called Garngad and was known as Garngad until well into the 20th century. This name change to Royston was put into effect in 1942, the new name being simply a shortened version of nearby Robroyston.

Ruchazie

The district of Ruchazie (Rachesie, Rachaisie or Roughhazy in older spellings) was a small township of some importance back in the 16th, 17th and 18th centuries. Set high on the hill overlooking the lands that swept down to the old Glasgow city centre growing up around its cathedral, this was a place whose inhabitants profited from some of the best agricultural land in the district and where, it is said, royalty visiting the Glasgow bishops at their country lodge at Provanhall would stay a while to enjoy the local hunting that this pleasant countryside offered. The name Ruchazie most probably goes back several centuries before those days to older Gaelic words: an earlier *rath* [ra] or fort built close to the *easaidh* [essee] which was the water cascading downwards in cataracts and small waterfalls out of the Hogganfield Loch into the Molendinar Burn that flowed finally into the Clyde far below. An ideal site up there for a fortified settlement.

Ruchill

A standing theory is that this comes from Gaelic *ruadh coille* [roo-uh kuhl-yuh] which means red wood. But Gaelic would normally have it the other way around, saying wood red. The truth may well be less romantic and less in need of a connection with the Gaelic. In the 17th century the area was recorded as Roughhill. And when in 1892 a public park was laid out here the poor quality of the soil was commented on. The Ruch– is most likely to be Scots *ruch,* which is rough.

RUTHERGLEN

A traditional account is that this is Gaelic *ruadh gleann* [roo-uhG gl-yown] meaning red glen. But this is hard to reconcile with certain facts. Even it were plausible to describe some vegetation or soil as reddish, there is no glen here. The form of the name has varied. Although a record dated 1153 has the spelling Ruthirglen, there are two 17th century maps (Blaeu and Pont) which give the town as Ruglan. If this latter was the way people actually pronounced the name it could perhaps indicate a Gaelic sort of omission of the –th– in speech.

The Ruglan version could tempt us into rather wild conjectures. Gaelic *rìgh-lann* [ree-lown] would be a royal residence. Cumbric (pre-dating any Gaelic) *rhyg lann* would be an enclosure for rye (a parallel to Auchenshuggle across the river).

But both these accounts fail to satisfy the –uther– element of the name, which has been present since the 12th century and so cannot just be ignored.

A more plausible approach to the name of Rutherglen is to make the word split not *before* the g (–glan or –glen) but *after* the g (–lan or –len). We could then begin to look at the name in connection with some centuries of previous history. The fact that Rutherglen was granted status as a royal borough by King David I in the year 1126 implies that the township had been in existence for a long time prior to that and had grown to become a place of some importance, situated up near the end of the tidal reach of the Clyde.

An account which does do some justice to the –uther– is based on the name of an older king of Strathclyde, called Rhydderch, who ruled around 580 to 614, based most probably down the river at Dumbarton. The word split is Rhydderch/lann. The –dd– in this Cumbric name, going by Welsh, would be pronounced –th–; the –ch as in Scottish loch. A Cumbric *lann* could be an enclosure or often the land around a church. Perhaps the place was named after this royal personage, in years some time after the Christianizing work of Ninian and Kentigern.

St Enoch

Who is the saint remembered these days for the underground station and shopping mall in that name? Enoch is a masculine name but in the 6th century St Enoch was a woman, a woman known by several names (or at least written about under several names), including Teneu, Thenava and Thanea. She was the mother of St Kentigern, also known as St Mungo. And her son Mungo established a church in Glasgow, which is now Glasgow cathedral.

The elegant St Enoch's Square as it was right at the end of the 18th century. The expanding city needed a new church and the site of an ancient chapel of St Thenew (another name for St Enoch) was chosen here in the city's then west end.

*St Enoch's Square as it was around 1910. The first
St Enoch's Church had soon become too small to
accommodate its growing congregation and this new
church replaced it in 1827 and survived until 1925. The
vast St Enoch's Hotel on the left, forming the entrance
to the railway station called St Enoch's, was knocked
down in 1977 to make way for a shopping mall, the
station having already closed in 1966. The little building
in the centre is the entrance to Glasgow's St Enoch
underground station. It's still there, but has changed
into a café.*

St Rollox

No longer a district but a name still retained in some proper names (a railway yard, a retail park, a hair salon) this saint's name was originally St Roch, an early medieval saint from Montpellier whose reputation for curing the sick made him a saviour for victims of the plague and a patron saint of many categories of the afflicted, including the falsely accused. His name occurs, more recognizably, in the Glaswegian schools called St Roch's. The less recognizable Rollox is popularly said to be a mangled form of Roch's Loch.

Sauchiehall St

The name of this famous street has nothing to do with any hall. The –hall element (which has been there for over 200 years) is from a Scots word *haugh* which is standardly defined as a waterside meadow or fertile land beside a river. The Sauchie– is also Scots in origin. It comes from *saugh* which is Scots for willow tree. So this was the willowy meadow down by the riverside. Well, that is what the name should mean: willows and a meadow there no doubt were, but nearness to the riverside? The Clyde is (and was) clearly too far away. So is the *haugh* a fiction? No, a map of Glasgow drawn around the end of the 18th century by Peter Fleming shows the street as Saughyhall Road and just south of it, stretching along the terminations of the newly laid out Bath Street and St Vincent Street, is a wooded area, a park, called Willow

Bank in which there are two little elongated ponds, creating enough of a *haugh* for the road passing it by to be named accordingly.

Shawlands

Shaw is Scots for wood and is related to Old English *sceaga* similarly meaning wood. This is simply wooded land.

Shettleston

Glasgow cathedral records hold a reference to the 'home of the daughter of Sedin', the text occurring in a papal bull dated 1179. Sedin is not otherwise known. Frequent later references are made to Schedinestun, this being the *tún* or homestead or farm of a certain Schedin. The spelling of the name went through many variations (a far from unusual occurrence) until by the end of the 18th century it had settled down as Shuttleston, which has led some to suggest (perhaps jokingly) that the connection with the high level of weaving activity in this former village lying outside Glasgow was being picked up in the name.

Singer

Nowadays the name Singer is to be found only in a few street names and in the name of a railway station. As any local will tell you, there is no connection with song. The origin's commercial, economic. This was the site of the Scottish branch of the American sewing machine manufacturers, the Singer Company, which first set up here in

1884 and which grew and grew and could scarcely keep pace with rocketing demand and became a major employer, accounting at its peak for some 16,000 workplaces, until demand dropped and the plant was shut down in 1980 and the buildings finally flattened in 1998.

SPRINGBOIG

The name of the district comes from the name of a farm that was here once. *Boig* is a Scots variant on bog or boggy land. But why the Spring–? This was most likely the source of a burn, as in Springburn.

SPRINGBURN

No better account of this name can be given than that written by the late Minister of Springburn Parish Church, Frank Myers.

> It is difficult to realise that Springburn was once a beautiful village set among the hills carpeted with grass and crested with trees. The burn which gave the village its name was the overflow from a well of spring water rising in the fields near the west end of Mansel Street, flowing down what used to be called Knox's Open, eventually being named Union Street, running down the side of the derelict building which was once Woolworth's, under Springburn Road, past the Well Brae out to the fields beyond and so meandering until it finally ended in Possil Marsh. The burn coming from the spring was the only domestic supply of water.

STEPPS

This is from the Scots word *stepp* which means a wooden stave. Laid crossways, many *stepps* make a road, especially a road across boggy ground.

STOBCROSS

Stobcross is no longer a district but the name persists in street names and a quayside. In Scots a *stob* can be a tree stump or a wooden stake driven into the ground to mark a boundary. The −cross in this name could be either in the sense of a religious cross or a crossroads. A good number of placenames ending in −cross in Glasgow (and Scotland) refer to crossroads, but it is more likely that this was the site of a boundary-marking religious cross. The old lands around here used to belong to the Anderson family who acquired them in the late 16th century. They built a grand house here called Stobcross House, which was added to by subsequent owners and finally flattened in 1875 to allow for the construction of the Queen's Dock, but then with the vanishing of the need for a dock of such massive extent the land was again in-filled to become the site for today's Scottish Exhibition and Conference Centre, the SECC. Most Glaswegians these days will also associate Stob with Stobhill, the name of a hospital some miles further north.

Stobcross House used to be well outside the city boundaries.

Bell's Bridge, for pedestrians and cyclists, crosses the Clyde to the once rural site of the old Stobcross House, which is now home to the SECC complex. Seen here are the auditorium known as the Armadillo and the Crowne Plaza hotel.

STRATHBUNGO

The Strath– is the Gaelic *srath*, although mispronounced with the –th spoken, when in Gaelic it is silent [pronounced in Gaelic: stra]. A *srath* is a wide and flattish river valley. Which there isn't here, and there wasn't when the district first grew up, as a small village, in the early 18th century. And –bungo? The closest connection is to St Mungo, the 6th century bishop of Glasgow. The m and the b are quite easily blended into one. It seems most likely that this name was created for its emotive sound and Scottish connotations rather than as a description of position or purpose.

THORNLIEBANK

The first time the place appeared as a little village on a map was in the 18th century when it was recorded as Thorny Bank.

TORYGLEN

The original spelling as Torryglen, with double rr, more closely reflects the likely Gaelic origin of the name in *tòrr* (hill) with the ending –*aidh* [pronounced ee] to indicate place. But then the Gaelic word *tòrraidh* should really have come after, not before, the noun glen. An alternative idea that the north-eastern Scots word *torie*, which is the grub of a daddy-long-legs, should be involved is perhaps not so likely, unless the one-time owner of a farm here had emigrated south... Still 'hilly glen' does seem a touch contrived. Maybe this is one of those names that were made up for emotional or fanciful reasons rather than being based on surrounding reality.

TRONGATE

Not a gate but a street, the –gate coming from the old English word *geata* (street). And the Tron– is the old Scots name of a public weighing machine or the place where goods were weighed. The first weighing beam was in operation here in the middle of the 16th century, checking and taxing.

WHITEINCH

The –inch is from the Gaelic *innis* [pronounced: eensh] which means both island and water-meadow. Blaeu's 1654 map shows Whyt Inch actually as an island in the Clyde, in days before the river was widened and dredged. But why white? Possibly an old reference to the waters of the river that once broke over rocks and rippled here.

YOKER

This comes from the Gaelic *ìochdar* [ee-uhK-kuhr], the d not being pronounced, which means low-lying ground, which indeed it still is, making a natural site for a ferry across the Clyde.

What's in a Scottish Placename?

For a broader view of some 1200 placenames from the whole of Scotland

Lexus Books publishes

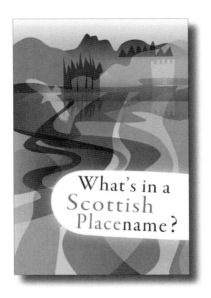

What's in a
Scottish
Placename?

which also delves into

origins and meanings

places and people

places and events

places and history

ISBN: 9781904737391